BEWARE
of the Food You Eat

DID YOU KNOW THAT:

• **Forty percent of communicable diseases are transmitted by food?**

• Food poisoning from salmonellosis in the United States has increased 1,000 percent since 1951?

• *Digestive upsets—more properly, food poisonings—affect more than 8,000,000 Americans annually, and are the second greatest cause of absence from school?*

• **There are 2,764 INTENTIONAL food additives?**

• Baby foods contain additives that are known to affect the supply of oxygen to the brain, while others may cause cancer?

• *Because of the high incidence of hog cholera and brucellosis in American pork and pork products, their importation has been forbidden by law in both England and Sweden?*

• Inspectors of the Pure Food and Drug Administration of the United States Department of Agriculture cannot possibly check more than 10 percent of the food destined for American consumption—and are able only to spot-check that amount?

• *Meat companies often scatter their processing and packaging plants in small and inaccessible areas to make it difficult for the pitifully inadequate forces of the USDA to inspect their facilities?*

• **The luxury super-supermarket you patronize may unwittingly be selling death in the form of packaged botulism—an acute form of food poisoning with a high mortality rate?**

SIGNET Titles of Special Interest

CONFESSIONS OF A SNEAKY ORGANIC COOK

Beware
of the Food You Eat

the revised, updated edition of
Poisons in Your Food

by Ruth Winter

with an Introduction by
Senator Walter F. Mondale

A SIGNET BOOK from
NEW AMERICAN LIBRARY
TIMES MIRROR

To Rose Rich Grosman

Library of Congress Catalog Card Number: 70-175032

This is an authorized reprint of a hardcover edition published by Crown Publishers, Inc. The hardcover edition was published simultaneously in Canada by General Publishing Company Limited.

 SIGNET TRADEMARK REG. U.S. PAT. OFF. AND FOREIGN COUNTRIES
REGISTERED TRADEMARK—MARCA REGISTRADA
HECHO EN CHICAGO, U.S.A.

SIGNET, SIGNET CLASSICS, SIGNETTE, MENTOR AND PLUME BOOKS *are published by The New American Library, Inc., 1301 Avenue of the Americas, New York, New York 10019*

FIRST PRINTING, JUNE, 1972

PRINTED IN THE UNITED STATES OF AMERICA

Acknowledgments

The author gratefully acknowledges permission to quote from their works by:

Jay M. Arena, M.D., Poison Control Center, Duke University Medical School, Durham, North Carolina, and the *Journal of the American Medical Association* for "Poisonings and Other Health Hazards Associated with Use of Detergents," *JAMA,* October 5, 1964.

William B. Deichmann, Ph.D., Professor and Chairman of the Department of Pharmacology and the Research and Teaching Center of Toxicology, University of Miami, Coral Gables, Florida, for "Retention of Pesticides in Human Adipose Tissues—Preliminary Report," *Industrial Medicine and Surgery,* 3713, March, 1968.

Ben Feingold, M.D., Chief, Department of Allergy, Kaiser Foundation Hospitals, "The Recognition of Food Additives in Allergic Diseases," presented before the American College of Allergists, Denver, Colorado, March 27, 1968.

Stephen D. Lockey, M.D., 60 North West End Avenue, Lancaster, Pennsylvania, for case histories.

Charles C. Thomas, Fort Lauderdale, Florida, publishers of *Chemical Carcinogenesis and Cancers,* by W. C. Hueper and W. D. Conway, for extracts from that volume.

The author gratefully acknowledges Representative Joseph G. Minish (Democrat-New Jersey) for his generous help.

Contents

Introduction

Like poverty and hunger before it, the issue of health hazards in food is a matter discussed more by experts than by the consuming public. Like those other issues as well, the issue of health hazards is gradually commanding more attention through the pioneering efforts of writers like Ruth Winter.

Our knowledge about health hazards in food is like the tip of an iceberg. This book touches on much of the submerged knowledge.

Testimony on meat, poultry and egg inspection legislation called attention to significant hazards, and information on fish clearly shows the need for legislative action here as well.

But protein products are not the only problem. Mrs. Winter graphically summarizes information about agricultural chemicals, food additives, and water pollutants. And in her discussion of the potential problems of catering, vending machines, convenience foods, and home freezing, she brings forward information that has not been widely discussed before.

Beware of the Food You Eat is a compendium of information for consumers. Mrs. Winter has presented scientific literature in an eminently readable form, giving laymen a sense of the nature of the controversy and, by extension, an opportunity to participate in its resolution.

For housewives and husbands, industry representatives, legislators, public officials, and scientists are all consumers of the food Mrs. Winter is writing about in this volume. It will take the efforts of all of us to mount the campaign of action suggested in Chapter 10 of this book. Meat, poultry and egg legislation passed in recent years is solid testimony to the effectiveness of the aroused citizen.

Behind the specifics is the underlying issue of the interaction of technology and society. Modern techniques of growing and processing the food we eat are remarkable. Today we may be on the verge of preventing worldwide starvation through this technology.

But *Beware of the Food You Eat* presents graphic evidence that this technology creates the possibility of detrimental effects as well. Tipping the balance toward human welfare requires genuine concern from all whose lives are touched.

That means those of us who produce and process must be concerned with the human impact of our practices. Those of us who use the products also have to defend ourselves.

All of us need information and the determination to force consideration of health and safety among those who seem to be unwilling to take such considerations upon themselves, whether in producing, processing, or consuming publics.

This book provides a good deal of the information. It should also provoke the determination.

WALTER F. MONDALE

1

Have You Had Your Poisons Today?

When you sit down to eat a meal in your own home or in one of the 541,000 eating and drinking places in the United States, you take it for granted that the food is wholesome and nutritious. Unless you have access to special knowledge—medical data, reports of qualified investigators, and findings behind the "vital statistics"—it may not occur to you that it is now impossible for you to eat an ordinary meal, or to give a child milk from a bottle, can, carton, or breast, that is not contaminated with pesticides.[1] The sausages, ham, hamburgers, or hot dogs you eat can be filled with hog blood, cereal, lungs, niacin, water, detergents, and/or sodium sulfite.[2] Even a baby's diet can be dangerous.

The bread you put in your mouth daily is literally embalmed to keep it feeling soft and fresh long after it isn't,[3] and the dessert you eat may be colored with a cancer-causing agent.[4]

The candy you put in the mouths of babies may not only be bad for their teeth but also dangerous because of contaminants such as wood splinters or rodent urine.[5]

"Impossible!" you are probably saying to yourself. "Our food is inspected and protected by law!"

But do you know that most foods from catering establishments, intrastate processing plants, and vending machines are almost never inspected?[6]

Do you know that only 10 percent of the foodstuffs adulterated with pesticides and chemicals or contaminated by insect filth and disease organisms is condemned by

11

Food and Drug inspectors? The other 90 percent gets through to unwary consumers.[7]

Do you know that approximately 60 percent of all fish eaten in the United States today originates in 116 foreign countries, a number of which have poor sanitation and living conditions? And that only 5 percent of such imports are inspected by the U.S. Food and Drug Administration? Are you aware that in the United States there are 4,200 fish-processing plants and 75,000 fishing vessels and only 62 inspectors from the U.S. Bureau of Fisheries? These inspectors check only 40 fish-processing plants, and then at the request and expense of the processors. To inspect adequately American fish products it was brought out at the 1969 Senate hearings on fish inspection that 6,273 trained persons would be required.[8]

The majority of food and drug inspectors are hard-working, grossly underpaid public servants. But despite the fact that most contamination and adulteration cannot be recognized without laboratory tests, 41 percent of the state milk and dairy supervisors and 71 percent of the state food inspection supervisors have no degrees.[9]

When I first wrote *Poisons in Your Food* in 1969, there were 800 U.S. Food and Drug Administration inspectors for the more than 100,000 food, drug, and cosmetic plants in the United States. Today, there are 649 inspectors, including 91 supervisors and 33 import inspectors. They are responsible for checking the food, drug, and cosmetic industries doing a business in excess of $130 billion a year.[10]

Many eminent physicians, university scientists, politicians, and industrialists contend that the American food supply is wholesome, safe, and nutritious and that it is abundant. Compared to what the starving peoples of the world must eat, and in contrast to the typhoid- or tuberculosis-contaminated food of the past, perhaps we should not complain. It is true that our food is abundant to the point that our taxes pay farmers not to grow crops. Our market shelves contain 8,000 items compared to the 1,500 they displayed before World War II. But this very increase in new items has created new dangers.

The National Academy of Sciences' Food Protection Committee warned in 1964: "Food scientists in industry and government are concerned about the increasing disparity between the rate of technological change in certain segments of the food industries and the level of efforts being made to evaluate and control health hazards associated with new products and processes."

Howard Bauman, Ph.D., vice president of the Pillsbury Company, warned of a "mass catastrophe" in the U.S. food supply at the American Health Association-sponsored National Conference on Food Protection in Denver, April 4–8, 1971.

"Can you imagine the runways and control towers of the thirties trying to keep track of and land jets as well as our airports today? It seems ridiculous, but that's exactly what we're doing in the food business," Dr. Bauman said.

FDA officials estimate that food poisoning in America has increased more than 1,000 percent since 1951, but no one really knows the true extent of the problem. Just one type of microbial foodborne illness, salmonella, attacks millions of Americans at an estimated cost of $300 million a year. Yet every year 10 or more states submit no reports about foodborne illnesses and many transmit only one or two reports in an entire year. In England and Wales, where food poisoning surveillance is highly developed, there were 792 outbreaks reported in 1968 compared to only 345 reported to the U.S. Center of Disease Control for the same period.[11]

The National Health Surveys have reported that digestive disturbances affect an estimated eight million Americans a year and that stomach upsets are the second leading cause of school absences. In fact, 42 percent of all illnesses suffered by man are foodborne.[12]

Even more dangerous than the obvious upset stomachs from food are the hidden illnesses, the unsuspected allergies, the irreparable damage to the cells of the unborn, and the slow poisons that are suspected of being a cause of cancer.

In the November 22, 1968, issue of *Science*, researchers from the University of Pittsburgh School of Medicine reported that a substance which causes liver cancer remained in the multiplying cancerous liver cells of rats "weeks to months" after it was removed from the rodent's diets. "For many different tissues," the Pittsburgh researchers said, "the required exposure time to a variety of carcinogens is considerably less than the shortest interval in which malignant cells can be detected with current techniques."

The National Academy of Sciences, the semiofficial adviser of the Food and Drug Administration, points out that "the number of people with tumors of any type at any one time in the United States approximates 3 per 1,000 population. Most tumors occur in older individuals who have been at risk for extended periods. . . . In many, if not most, cases the carcinogenic potential of a chemical is assessed using the same animals employed for assessing other toxic hazards. Carcinogenicity deserves special treatment, however, because of the frequently long induction periods required before tumors can be detected. . . ."[13]

A man who is not alone in his warnings, Dr. W. C. Hueper, former Chief of the Environmental Cancer Section, National Cancer Institute, said: "Apart from the fact that most of the food, drug and cosmetic dyes used in this and other countries have not adequately and competently been studied for possible carcinogenic properties in experimental animals, no pertinent reliable and valid information has ever been published concerning such possibilities by surveying occupational groups and other specially exposed population groups for carcinogenic effects from these agents when present in a considerably more impure state than that required for food dyes. Such investigations, however, are urgently needed because claims have been advanced, based on epidemiologically deficient evidence, that the use of butter yellow in Austria and in some oriental countries for coloring foodstuffs was responsible for an increased or excessive death rate from primary liver cancer among the populations consuming such foods."[14]

A fine example of our imperfect food-protection system was in the continued use of cyclamates, the artificial sweetener. As far back as 1958, the FDA was warned by its own scientists that cyclamate caused a significantly high rate of cancer in animals. In 1969, when it was finally banned, 175 million Americans—from kids chewing bubble gum to adults fighting fat—were swallowing cyclamates in large doses, although it was originally intended for diabetics and dieters under medical supervision.

Dr. Chauncey Leake, past president of the American Association for the Advancement of Science, and one of the nation's most distinguished pharmacologists, warned in 1963 that general use of new chemicals in large quantities has created a new hazard—subclinical poisoning—so insidious that physicians cannot connect the poison with the ailment.[15]

Rachel Carson showed in 1962 how fruit flies, the classic subject of genetics experiments, developed damaging and even fatal mutations when exposed to one of the common herbicides or to urethane. Urethane belongs to the group of chemicals called carbamates, from which an increasing number of insecticides and other agricultural chemicals are drawn.

"Two of the carbamates," she wrote, "are actually used to prevent sprouting of potatoes in storage—precisely because of their proven effect on stopping cell division."

Did Rachel Carson, a highly respected marine biologist, write her book in vain?

Yet, in "Protecting Our Foods," *The Yearbook of Agriculture,* 1966, the United States Department of Agriculture states on page 28: "Both Malathion and carbaryl are toxic to many insect pests. Unlike the chlorinated hydrocarbon materials they do not accumulate in tissues. Thus they can be applied up to or within a day or so of harvest on many human food and forage crops." (Carbaryl, like urethane, is a carbamate.) That Malathion has been approved for direct application to grain and to peanuts in the shell is described on page 146 of the same publication.

Until very recently, the ability of chemicals in food to cause mutations in human cells was not even considered. The National Academy of Sciences admits "tests for mutagenic action in mammals are in an early state of development . . . It is obvious that improved methods are needed for assessing mutagenesis . . . Mutagenesis differs from the other categories of embryotoxicity in that the defect, if not lethal, has the potential for reappearing in generations succeeding that in which it is first seen. Chemicals may alter not only fertilization and intrauterine development of fertilized ova but also the hereditary material itself [mutagenesis.]"[15a]

Dr. James Watson, who won the Nobel Prize for his work in genetics, told me at a science writers' conference in New York in 1971: "Americans are very casual about what they do to their genes." He pointed out that we carelessly introduce chemicals into our environment without knowing their long-term effect.

Until the past few years, scientists believed that the placenta surrounding the fetus protected the unborn child from toxic chemicals. It is now evident that not only can chemicals penetrate the placenta and affect the unborn child, but that the chemicals may become more concentrated and more toxic. All children born in the United States today have traces of DDT and other chlorinated hydrocarbon insecticides in their tissues.[15b]

Because of public pressure—mostly from a concern for wildlife—the use of DDT has been restricted, and it may eventually be outlawed.

But no matter how much the residues are reduced or how many noxious chemicals are recognized and banned, for those of us already subjected to as many as ten pesticides with every meal, it is too late. It may even be too late for those children we may someday have. Our tissues have been storing a generous supply of pesticides.[16]

The human digestive system is in itself a chemical engine. As it moves food through its conveyor belt, it adds digestive juices and secretions from the gallbladder, liver, and pancreas. The human body is a wondrous thing. It can detoxify many poisons through its liver and kidneys. It can fend off many germs, and even cancer—that is, if it is young and healthy. But what about infants and the elderly, whose resistance is not great? What about those with damaged kidneys and livers? allergies? chronic illnesses?

The use of chemicals in foods has soared from 419 million pounds in 1955 to more than 800 million today. Each of us eats more than three pounds of food additives a year.[17]

Added to the intentional and unintentional chemicals in our foods are the chemicals we ingest as medicines. Americans are the most medicated people in the world. Every year we swallow 37 billion doses of therapeutic pills, powders, capsules, and elixirs.[18]

In addition to the foods we eat and the medicines we take, we fill the sky with 130 million tons of noxious chemicals . . . carbon monoxides, hydrocarbons, nitrogen oxides, sulfur oxides, and particulates. Persons living in New York may inhale the equivalent of 730 pounds of chemicals a year.[19]

In addition to chemical purifiers, water today contains all sorts of pollutants, from detergents to sewage and industrial wastes.

Added to all the chemicals in food, water, medicine and air, a large percentage of the population smokes and/or drinks alcoholic beverages. Tobacco smoke may be an additional cancer-causing agent that breaks down the body's immunity, while alcohol may be a solvent that aids another noxious chemical.

Perhaps the many eminent, conservative scientists who warn us that we may be irrevocably poisoning ourselves are alarmists. On the other hand, before we draw that conclusion, we must rule out such things as the role of pesticides in infant mortality.

At least 25 percent of the conceptions in the United States end in intrauterine death.[20] Is it a coincidence that traces of pesticides have been found in the tissues of stillborn and unborn human babies in the United States? In some cases the concentration of pesticides in the fetus was almost as high as those of the mother or of young adults of comparable age.[21]

The question of why millions of Americans have ulcers, gallstones, kidney stones, and cancer must be answered. Although heredity is certainly implicated, so is diet.

We must ask ourselves why trichinosis is higher in the United States than in European countries. Why the American export of oranges has been endangered because the European Economic Market set its permissible biphenyl (decay retardant) residue rate 40 percent lower than the 110 parts per million rate set by the Federal Food and Drug Administration.[22]

We must ask ourselves if we must accept the suffering and loss of millions of hours from school and work caused by preventable intestinal upsets.

We hear much today of ecology; we worry about changes in wildlife when the "balance of nature" is upset. Yet one of the world's leading gastroenterologists, Dr. Burrill B. Crohn of New York, told his colleagues at the Eighth International Congress of Gastroenterology in Prague, in 1968, that the current high incidence of intestinal diseases may reflect a "disturbed ecology of the human race."

Dr. Crohn, who is Professor Emeritus of Medicine at Mount Sinai School of Medicine, said that he could not accept the view that the "upsurge" of these disorders could be laid to the stresses of the twentieth century. Other centuries have also been marked by stress, he said.

"Is the causation of these diseases to be found in the food we eat?" Dr. Crohn asked. "In the last decades of our ecology, our foods contain multiple chemical preservatives; our growing crops are sprayed with insecticides. We ingest multiple new drugs never before used, such as the coal-tar products of which aspirin and its derivatives are regular household remedies. The pollution of our drinking water might also be a factor."[23]

1 NOTES

1. R. E. Duggan and Keith Dawson, "Pesticides," a report on residues in food, U.S. Food and Drug Administration reports, *FDA Papers,* June, 1967, Vol. 1, No. 5. Tom Bogaard, chemist and general manager, Kenya Pyrethrum Co., interview with author, January, 1968. Dr. George Kupchik, Director of Environmental Health, American Public Health Association, tape-recorded interview with author, February 26, 1968.

2. William E. Jennings, director of the Division of Meat Inspection of New York, April 20, 1963, issue of *The National Provisioner.* Rodney E. Leonard, Deputy Assistant Secretary, Consumer and Marketing Service, U.S. Department of Agriculture, November 9, 1967, Congressional Hearings on Meat Inspection, Washington, D.C.

3. Howard J. Sanders, "Food Additives," *Chemical and Engineering News,* October 17, 1966. *Changing Times,* May, 1960.

4. W. C. Hueper, former chief of the Environmental Section, National Cancer Institute, and W. D. Conway, *Chemical Carcinogenesis and Cancers* (Springfield, Ill.: C. C. Thomas, 1965), p. 645.

5. Chocolate mint candy, Utah, packed under insanitary conditions; contains wood splinters, cloth fibers, metal fragments, *FDA Papers,* November, 1967, p. 36. Peppermint candy, South Carolina and Texas, on December 20, 1966, had rodent contamination, *FDA Papers,* April, 1967 p. 31.

6. Dr. George Kupchik, taped interview February 26, 1968.

7. *Ibid.*

8. James Brooker, Director of the Division of Fisheries Inspection, personal communication with author, February,

1971; H. R. Robinson, Chairman, Legislative Committee, American Shrimp Canners Association, New Orleans, U.S. Senate Hearings, Wholesale Fish and Fishery Products Act of 1969, July 1, 2, and 14, 1969.

9. "A Study of State and Local Food and Drug Programs," report to the Commissioner, Food and Drug Administration, U.S. Department of Health, Education and Welfare, February, 1965.

10. U.S. Food and Drug Administration, Washington, D.C., personal communication with author, February, 1971.

11. National Conference on Food Protection, Denver, Colorado, April 4–8, 1971, preliminary position papers, p. 119.

12. Robert W. Harkins, Ph.D., Chicago, *Journal of the American Medical Association*, November 6, 1967, Vol. 202, No. 6.

13. "Evaluating the Safety of Food Chemicals," National Academy of Sciences, Washington, D.C., report, 1970.

14. Hueper and Conway, *op. cit.*, p. 645.

15. *Medical World News*, January, 1963.

15a. "Evaluating the Safety of Food Chemicals," National Academy of Sciences, Washington, D.C., 1970, p. 28.

15b. M. R. Zavon, R. Tye, and L. Latorre, The Kettering Laboratory, Department of Environmental Health, College of Medicine, University of Cincinnati, Cincinnati, Ohio, "Chlorinated Hydrocarbon Insecticide Content of the Neonate," *Annals of the New York Academy of Sciences*, June 23, 1969, Volume 160, pp. 196–200.

16. "Pesticide Residues in Food," Food and Agriculture Organization of the United Nations, 1967, p. 15.

17. Howard J. Sanders, "Food Additives," *Chemical and Engineering News*, October 17, 1966. James L. Goddard, M.D., FDA Commissioner, tape-recorded interview with author, May, 1968.

18. Pharmaceutical Manufacturers Association, Washington, D.C., fact booklet, 1967.

19. Essex County, New Jersey, Medical Society bulletin.

20. David Zimmermann, "Unmasking the Agents That Can Make Genes Go Wrong," *Medical World News*, May 17, 1968.

21. Harold Schmeck, "Pesticides Found in Unborn Babies," *New York Times*, Sunday, January 30, 1966.

22. United States Department of Agriculture news release, "USDA Studies Residue Control of Biphenyl-Treated Oranges," October 15, 1968.

23. *Medical Tribune*, August 8, 1968.

2

How to Kill Insects— and People

It was a warm June day—warm enough for a young mother to keep the window open while she gave her five-year-old son a bath. Insecticide that was being used to protect the elm trees in front of the house was accidentally whooshed through the window, filling the bathroom with fumes. Both the mother and the little boy choked and coughed for fifteen minutes.

Eight months later, the child was taken to the family physician because a skin wound failed to heal properly. Worried about the slow healing and the boy's loss of appetite, listlessness, and enlarged spleen, the doctor had him admitted to the hospital for further tests. An examination of the child's blood marrow confirmed the dread diagnosis—acute granulocytic leukemia.

The child was treated by various means—the latest drugs, X rays, blood transfusions—but as spring was just beginning to fill the air and the elm trees in front of the house were once again proudly displaying their buds, the little boy died.

In the following February his mother was referred to the Mayo Clinic because of a suspiciously low white blood count that had first been noticed three months earlier, when she suffered an attack of the "flu" with a fever of 103 degrees. Her fever had cleared and she had felt well—that is, as well as a mother who has lost a child can ever feel. At the hospital, an examination of her bone marrow revealed acute leukemia. She was put on drugs, and received twenty-eight blood transfusions. But, nine months after her little boy died, she also succumbed to leukemia.[1]

High in the Andes of Colombia there is a town called Chiquinquirá, about which almost no one had heard until a day in November, 1967. Chiquinquirá, like farm towns everywhere, rose early for breakfast that particular morning. By 8:30 A.M., the first child was dead.

By afternoon, more than 130 persons, most of them children, violently sick and near death, were brought to the hospital.

What was the deadly poison? About a pint of the pesticide parathion. On the hundred-mile truck journey from Bogotá, a parathion container broke in its carton, and the insecticide leaked into the bags of flour. A baker used two of the bags of flour for his early-morning batch of bread. Within four hours, the first of the victims had died.[2]

Just a month earlier, the story had been the same in Tijuana, Mexico. There, seventeen died from bread contaminated with parathion.[3]

Parathion poisoning has occurred within the United States, although not as yet on such a mass scale. Shortly after the news of the Colombian deaths broke, the Health Department of New York warned that parathion was being sold by itinerant peddlers in the city as an insecticide. Health Commissioner Edward O'Rourke said that an eighteen-month-old child had swallowed the poison, which had been sold as a vermin exterminator in an unlabeled baby-food jar.[4]

Parathion is also used as an agricultural spray, particularly on vegetables and cotton. In its September, 1968, report, the Food and Drug Administration noted that 340 crates of celery had been seized in Florida because of contamination with parathion residue.

The awful question that haunts many scientists today is whether the one to a dozen pesticides we may ingest with every meal may cause or contribute to a slow poisoning of our systems and those of our unborn children.

The American Medical Association Council on Drugs, which maintains a registry on blood dyscrasias, has a list of eighteen reports of major blood abnormalities associated with exposure to an insecticide, lindane, or benzene hexachloride (BHC). Dozens of other cases linking blood disorders and insecticides have also been reported.[5]

Since 1957, five Californians have died from aplastic anemia or related blood dyscrasias; each of them had been exposed to the chlorinated hydrocarbon pesticide known as lindane. None of the deaths was listed as due to lindane in the mortality statistics.[6] These deaths, which came to the attention of the California State Health Department, included that of a thirty-nine-year-old wife of a physician. The interior of her home had been treated, every three months for several years, by a pest-control operator who used vaporizing lindane. On one occasion, an insect was found just after treatment, and the application was repeated. A heavier than usual "white film" was found over the walls and surfaces of the home. After cleaning the house thoroughly, the housewife became seriously ill, and a diagnosis of aplastic anemia was made. On several occasions before her death, she noted relapses after eating in a restaurant where a lindane vaporizer was in operation.

Dr. Irma West, of the Bureau of Occupational Health, State of California Department of Public Health, said before a Senate committee in 1963: "With the notable exception of the organic phosphate pesticides, physicians do not have laboratory tests available to confirm diagnoses of agricultural chemical poisoning. Diagnoses can be made on a clinical or circumstantial basis only. Signs and symptoms may be similar to many other conditions. Just about any illness can be attributed to chemicals and there is little help in confirming, refuting or clarifying the situation. Bona fide cases can be buried among the erroneous, and vice versa.

"There are instances of the most astute diagnostic skill and examples of the most dismal ignorance in the medical handling of pesticide poisoning."

She gave as an example the case of a young boy who had been applying an organic phosphate pesticide all day. He developed symptoms of poisoning. He was sent home twice in the course of several hours by the same physician—once with a tranquilizer for medication.

"The second time he came for medical help," Dr. West said, "the label from the chemical was presented to the physician, who sent him home again with several atropine pills, a first aid recommendation on some labels but an entirely inadequate treatment. The boy died during the night at home, still in his contaminated work clothes."

Rachel Carson, in her best seller *Silent Spring*,[7] made

the world aware of the potential dangers of pesticides. For an all-too-brief period, there was a great public outcry for stricter laws and controls. Then reassuring books and articles flooded the market—most of them sponsored by pesticide manufacturers. Rachel Carson was accused of being a sensationalist, and the public settled back into apathy.

In one year, United States farmers use an estimated 175,826,000 pounds of insecticides, averaging $250 worth per farm.[8] There are 500 chemical compounds used in more than 54,000 formulations for pest control alone.[9] Total pesticide sales were up 37 percent between 1964 and 1966. Domestic use of pesticides was up an estimated 15 percent in 1968 over 1967.[10]

Pesticides certainly have proved their economic worth. In the United States, some 10,000 species of insects are classed as public enemies. Of these, several hundred species are particularly destructive, and require some degree of control. The estimated cost of insect-control measures to protect our food and fibers is $704 million a year.

"Without insecticides," according to Agricultural Research Division entomologist C. H. Hoffman and Chief of the Stored Product Insect Research Branch L. S. Henderson, both of the United States Department of Agriculture, "production of livestock would soon drop about 25 percent and production of crops about 30 percent. Food prices might then go up as much as 50 to 75 percent and the food still not be of high quality."[11]

The fact that farmers are producing more than they can sell, killing beef and dumping milk to raise prices, is beside the point.[12] The real question is whether the price we are paying for insecticides may be cancer, birth defects, liver diseases, and even mental illness.

The age of organic pesticides began with DDT, still the most widely known and used pest killer. It combines chlorine with carbon and hydrogen. It is manufactured from chlorine (a deadly gas), benzene (a leukemia-causing agent), and alcohol (toxic to the liver), which react and combine to form the DDT molecule.

DDT was first described in 1874 by the German chemist Othmar Zeidler. Its insecticidal value was recognized in 1939 by Paul Mueller, a scientist working for the Geigy Dye Company in Basel, Switzerland. The Swiss potato crop was threatened that year by the Colorado potato beetle, and the company gave Dr. Mueller a sample of DDT for testing. The beetle was exterminated, and the value of DDT against other insecticides was heralded. Dr. Mueller's work won him the Nobel Prize in 1948.

The Geigy Company shipped samples of DDT to the United States for testing in 1942. The good results soon led to the importing of the insecticide. At first it was used only for the Armed Forces, but after World War II it was unleashed on the general market. DDT can now be found in the bodies of everyone reading this book.[13]

Penguins in Antarctica, elk on the high slopes of North America, oysters in Chesapeake Bay, and the owl in your attic—all now have DDT or traces of some other lethal pesticide in their systems. Ducks, bald and golden eagles, deer, perch, and shellfish—in fact, probably every living creature on this earth—even those separated from farmers by thousands of miles, by depths of oceans, by mountains and clouds, have pesticides in their flesh.

Pesticides used to control harmful insects are washed into streams or lakes, where they are absorbed by tiny plankton, which are eaten by small fish, which are eaten by bigger fish, which are eaten by birds, which in turn may be eaten by other creatures.

For example, E. G. Hunt, of the California Department of Fish and Game, Sacramento, described startling results at Clearlake, California, where pesticides were used to control gnats. Residue levels of the pesticides in the lake area after thirteen months showed 10 parts per million in plankton, 903 parts per million in smaller fish, 2,690 parts per million in carnivorous fish, and 2,134 parts per million in the fat of fish-eating birds. This represented nearly a 100,000-fold increase of harmful chemicals in fish-eating birds.[14]

Researchers at the University of Florida College of Medicine, Gainesville, Florida, reported in 1971 that their experiments with dogfish, a member of the shark family, show just how persistent DDT can be. They fed DDT to the fish and found the chemical concentrated mainly in their fat and liver. Unlike mammals, the fish did not excrete the chemical. When the fish died, all the DDT returned unchanged to the sea to perpetuate another cycle.[14a]

Except for a few cannibals in remote parts of the world (they too probably have pesticides in their flesh), man is at the end of the food chain. The dose of pesticide he may receive from eating fish that have eaten smaller fish that have eaten plankton or from game and beef that have eaten insecticide-contaminated feed defies measurement. That is, except for rare instances, no one even tries to measure it.[15] DDT is toxic to man and beast. How toxic and in what dosage is unknown.[16]

"Although DDT is the most widely used insecticide, its mechanism of action remains uncertain," said Cornell University researchers R. D. O'Brien and F. Matsumara in the General Medical Science's United States Health, Education, and Welfare Department report for 1964–1965. "However, the symptoms of DDT poisoning in animals—which include tremors, convulsions and paralysis—indicate that the poison attacks the central nervous system."

There is no doubt that DDT can cause deaths—the *Journal of the American Medical Association* in 1951 listed twenty-three cases of DDT ingestion, four of which resulted in death.

California, which uses 20 percent of the nation's pesticides, has about 1,000 cases of occupational diseases associated with exposure to them or other agricultural chemicals annually. There are about 150 pesticide deaths a year in the nation. How many children are poisoned by ingestion of pesticides can be gleaned from the fact that in California alone, 3,000 youngsters a year are treated in hospital emergency rooms for pesticide poisoning.[17]

Take the case of a three-year-old California girl who in 1965 was playing near where her parents were picking berries on a farm. She took the cap off a gallon can of 40 percent tetraethyl pyrophosphate, TEPP, a pesticide. She put her finger in the can and then sucked her finger. She immediately vomited, became unconscious, and was dead on arrival at the hospital.[18]

In the state of Washington, a small boy was poisoned almost fatally from parathion that had survived the winter snow and rain after being spilled on the driveway of his home. The youngster had eaten mud pies made from contaminated soil that was found to contain 1 percent parathion.[19]

An eighteen-month-old child of a California agricultural aircraft pilot was found at home in a state of acute respiratory distress, semiconscious, and with pinpoint pupils. In the hospital, she was placed in a resuscitator and treated by a skilled physician for severe organic phosphate poisoning, from which she recovered. On the morning of the illness, her father had come home after applying a highly toxic phosphate ester pesticide. He was reported to have cleaned his boots with paper towels and then thrown the towels into the wastebasket and placed his boots in the bathroom. The child touched either the boots or the paper in the wastebasket.[20]

The preceding cases concerned relatively large amounts of pesticides as poisons, but what of even smaller residues?

A decline in the learning ability of quail given DDT-doctored food was reported by scientists attending a recent American Institute of Biological Sciences meeting.[21] The researcher Dr. Douglas James, Associate Professor of Zoology at the University of Arkansas, said the amount of DDT the birds ate had been considered safe.

Evidently DDT and another class of pesticides, organophosphate compounds, have similar mental effects on man—the worst effects when given in combination. A study at the University of Colorado Medical Center shows that men exposed to these pesticides over a long period have relatively slow reaction times and poor memories, tend to lose their vitality and become unusually irritable.

Reporting at a New York Academy of Sciences Conference on "The Biological Effects of Pesticides in the Mammalian System," Dr. David R. Metcalf, Assistant Professor of Psychiatry and head of the Division of Electroencephalography at the University of Colorado, pointed out that there are few hard neurologic signs of poisoning, such as sensory or motor deficits, but that there is a significant increase in the soft signs, namely: "Persistent, easily demonstrated slowness or thinking and memory deficit."[22]

A San Francisco psychiatrist, Dr. Douglas Gordon Campbell, of the University of California Medical School, said in 1963 that in taking a history, every physician should ask if the patient uses sprays in the home, in the garden, or on the job."[23]

It has been known for a number of years that DDT and other chlorinated hydrocarbons interfere with the reproductive systems of birds, causing eggshell thinning and the near extinction of some species. In 1971, Dr. William L.

Heinrichs, of the University of Washington, told an American Cancer Society-sponsored meeting of science writers that rat studies he has done indicate DDT can interfere with mammalian reproduction, and that the increased incidence of ovarian cysts among young women may be a result of DDT exposure.[24]

According to Dr. Heinrichs, DDT induces in rats a syndrome that includes the induction of numerous ovarian cysts and failure of reproduction. In recent years there has been evidence of increase of a similar condition among young women, he said. He theorizes that DDT acted on the pituitary gland of the young women—perhaps prenatally—and altered the hormonal "programming."

If nothing else has stirred American men to action concerning the dangers of pesticides, then perhaps the work of Dr. Richard M. Welch, a biochemical pharmacologist, will. Dr. Welch, who works at the Burroughs-Wellcome Research Laboratories, Tuckahoe, New York, testified at state hearings in Wisconsin, on banning DDT, that the pesticide may seriously affect human sex organs.[24a]

Dr. Welch said experiments he conducted with rats showed alterations in the sexual mechanisms of both males and females.

"If one can extrapolate data from animals to man, then one can say this change in animals probably does occur in man," he observed.

The World Health Organization (WHO) noted in 1966 that both women and cows secrete DDT in their milk. The content of DDT in cows' milk has been found to be lower than in mothers' milk. At the hearings in Wisconsin in 1969, a Swedish toxicologist reported that breast-fed infants throughout the world were ingesting approximately twice the amount of DDT compounds recommended as a maximum daily intake by the World Health Organization.[24b] In 1971, a Japanese researcher, Dr. Takao Nishimoto, reported that mothers' milk in Japan contained up to 30 times the maximum dose of dieldrin that WHO considers safe. Dieldrin is a cousin of DDT. The Japanese researcher also detected an average of 0.08 ppm (parts per million) of DDT in the mothers' milk.[24c]

An experimental study was done some years ago on the effects on man of prolonged ingestion of small doses of DDT in the form of oily solution in capsules or emulsions in milk. Administration was continued for as long as eighteen months. The authors noted that the accumulation of the insecticide in the fatty tissues and the urinary excretion of its metabolite, DDA, were proportional to the dose of DDT ingested. A state of equilibrium was reached after about a year, and the concentration of DDT accumulated in the fatty tissues reached an average of 234 parts per million in subjects who had ingested 35 milligrams of DDT per day. No ill effects were noted.

However, in April, 1971, researchers at the University of Miami reported that DDT tissue retention rose suddenly and dramatically when another pesticide, aldrin, was added to the diet of dogs.[24d]

"At this time one can only speculate about the mechanism responsible for the sudden and unexpected rise in the concentrations of DDT and its metabolites [breakdown products] in the blood and fat when aldrin is added to the dietary intake," the Miami researchers wrote. Aldrin is a commonly used pesticide.

Donald S. Kwalick, M.D., Director of the New Jersey State Health Department's Community Study on Pesticides, wrote in the May, 1971, issue of the *Journal of the New Jersey Medical Society*: "The acute effects of pesticide poisonings are well understood but the effects of long-term, low-level exposure are unknown. To determine whether chronic effects exist, the New Jersey Community Study on Pesticides has selected 200 individuals (farmers, pest-control applicators and formulators) who are occupationally exposed to high levels of pesticides."[24e]

Dr. Kwalick said these people are being compared to 50 minimally exposed controls for any differences in body chemistry or in physical signs or symptoms.

The Pesticide Projects, according to Dr. Kwalick, has found that persons exposed to pesticides in their occupations had a residue of DDT of 28.8 ppb (parts per billion) whereas the minimally exposed population had a 4.4 ppb. He said the researchers have not yet found any abnormal biochemical chronic changes in the exposed individuals. However, they have discovered that the exposed group had "a statistically significant difference in the number of individuals with hearing and eye problems, chronic cough and sinusitis, dizziness, headaches, and high blood pressure."

No one knows the ultimate effects of DDT or of its

equally toxic cousins, dieldrin, aldrin, endrin, toxaphene, lindane, methoxychlor, and heptachlor. It is known that even in relatively minute amounts, dire results can occur. Take the case of the mysterious stain poisoning:

At 1:00 P.M. on a Tuesday in July, a twenty-seven-year-old woman (we'll call her Mrs. Anderson) walked into the medical department of a United States Naval Medical Research unit because of a "tender" lump on her forehead. At breakfast, apparently, she was in her usual fine state of health. At noon, when she complained of a headache and feeling ill, the patient's husband discovered the prominent lump on her forehead. Neither she nor her husband could explain the swelling.

Upon admission to the hospital, Mrs. Anderson appeared acutely ill, complaining of a headache and nausea. She walked with slight assistance into the examining room. She became easily confused by simple questioning about her morning activities. Within five minutes, she suddenly became speechless and had a convulsion that lasted approximately two minutes. She frothed at the mouth and was confused for two or three minutes after remission of the convulsion. The patient remained semiconscious for fifteen to twenty minutes. After awakening, she could not even remember coming to the naval facility or the subsequent events.

On that same July day, a neighbor of Mrs. Anderson, a Miss Baxter, had a friend, Mr. Cummings, to lunch. About one and a half hours later, Mr. Cummings was found on the floor of his home, "frothing at the mouth, jerking and beating his head on the floor. After a short, semiconscious period, he vomited several times and then felt much better."

At 5:00 P.M., Miss Baxter saw the servant of Mr. and Mrs. Anderson walk from the house, stop, and suddenly fall to the ground with jerking movements, frothing at the mouth. Following several minutes of semiconsciousness, she was taken to a hospital.

Like Sherlock Holmes, the Navy doctors meticulously tracked down the source of the poisoning. It turned out that a servant of the Andersons had made bread the previous evening. Part of the bread had been given by Mrs. Anderson to Miss Baxter, who in turn served it for lunch. Miss Baxter remained in good health despite the fact that she had consumed the same quantity of bread, one roll, as did her friend Mr. Cummings, who developed convulsions shortly after the meal. Mr. Anderson had consumed even larger quantities of bread than his wife on the morning of her seizures. He had no ill effects. How much bread the couple's children ate is unclear, but they ate large quantities of cookies made by the servant at the same time that she made the bread.

The Navy doctors fed part of the bread from the batch eaten by Mrs. Anderson to a rat and two dogs. The rat died within twelve hours. The dogs began to salivate within two hours, then had convulsions, but recovered.

Mr. Anderson recalled that a case containing twelve five-pound sacks of flour purchased in March had been stained. An investigation of flour from one of the stained sacks was made. Flour from the sack was fed to a dog. Within thirty minutes the dog began to salivate. Two hours later, the dog was dead.

The stain proved to be endrin. Ironically, when the blood, serum, and urine of all the persons involved were studied, eight pesticides were found. Why the others who ate the bread did not suffer the convulsions was not clear, but perhaps they were lucky to have missed the "stained" flour or they were resistant to the poison.[25]

In their article about the case in the *Journal of the American Medical Association,* the doctors said: "The toxicity of endrin to humans appears to involve stimulation of the central nervous system. Other than neurological damage secondary to inadequate ventilation during seizures, the only suggestions in the literature of permanent injury directly due to endrin are:

"1. A possible synergism [joint action] of endrin and DDT believed to be responsible for a peripheral neuropathy [nerve damage.]

"2. A patient whose course and electroencephalograms indicate brain stem injury.

"No specific findings at autopsy of humans or animals have been reported from acute poisoning, but chronic endrin poisoning may produce liver changes."

How toxic are endrin's cousins aldrin and dieldrin? It depends on where you look. For instance, quoting from the article "A Summary of Work on Aldrin and Dieldrin Toxicity at the Kettering Laboratory," by Dr. Frank P. Cleveland of Cincinnati, in the *Archives of Environmental Health,* 1966:

"It is my opinion that aldrin and dieldrin are neither carcinogenic nor tumorigenic. In a long-term experiment when aldrin and dieldrin were fed in various concentrations to rats, the incidence of tumors in the experimental animals was similar to that of control rats. . . .

"While these substances are highly toxic to insects and experimental animals, no definitive evidence is available that any harm has resulted to mankind. . . ."[26]

From the "Evaluation of Some Pesticide Residues in Food," published by the Food and Agriculture Organization of the United Nations World Health Organization, 1966:

"The primary site of action of aldrin and dieldrin is the central nervous system. CNS stimulation is the cause of death in acute poisoning. Signs of CNS stimulation are also seen after repeated high doses. Repeated doses at lower levels give rise to liver damage, and in this respect, young dogs are more susceptible than older dogs.

"In one long-term feeding experiment in rats there was a general increase in tumor production in the experimental animals at the lower dosage levels as compared to the controls, but the liver was not particularly affected. Liver tumors, however, significantly increased at dosage levels of 10 parts per million in one strain of mice susceptible to the development of these tumors."[27]

The World Health experts, while pointing out that more work is required to determine dieldrin and aldrin's effect on the liver, estimated the acceptable daily intake for aldrin and dieldrin combined in man was 0.0001 mg./kg. per day. It also pointed out that in one study, dieldrin was found in human mother's milk at an average concentration of .006 parts per million.

The danger of pesticides to unborn babies is just beginning to be recognized. Physicians from Baylor University and Texas Children's Hospital, writing in the *Journal of the American Medical Association* on December 18, 1967, on "Maternal Exposures to Potential Teratogens" (causes of birth defects), found that 125 mothers—52 percent of those studied—had been exposed to pesticides during pregnancy. Two infants with major malformations were born to mothers who had first trimester exposures to insecticides and two others to mothers who had later exposures.[28]

Israeli researchers reported in the February, 1970, issue

of *Archives of Environmental Health* that a study of 35 pregnant and 33 nonpregnant women revealed that in pregnancy the metabolism of insecticides is enhanced and that organochlorine insecticides pass the placental barrier.[28a]

In 1970, it was announced that a herbicide, 2,4,5–T, commonly used in Vietnam as a defoliant and in this country as a herbicide, caused birth deformities in mice and possibly in Vietnamese. On April 15, 1970, the Secretary of Agriculture, Clifford Hardin, restricted the use of 2,4,5–T for use around the home and for use in aquatic areas. He also restricted the manufacture of nonliquid formulations for use around the home and on food crops grown for human consumption. Most of us have been exposed, even though the chemical has been found to be highly capable of causing defective offspring in mammals.[29]

The National Cancer Institute is awarding hundreds of thousands of dollars to study the potential cancer-causing properties of old and new pesticides, including fungicides, herbicides, and growth regulators.

"Particular attention is being paid to synergists, compounds that enhance the action of pesticides. The use of synergists could decrease the amount of pesticide needed and thus reduce the problem of pesticide residue in crops," according to the National Advisory Cancer Council. "Synergists appear to act by increasing the toxicity of the pesticides for the insect. However, by a similar process they may also increase the toxicity of pesticides and other suspect cancer-causing chemicals for people. In one study, the effect of synergists, such as piperonyl compounds, on the metabolism of cancer-causing hydrocarbon compound in rats is being studied. Another project is investigating the hazard of the synergist piperonyl butoxide in the diet as associated with the other environmental chemicals—DDT and carcinogenic aromatic amine."[30] These studies are long-term, and no definitive results have been reported.

However, one former government scientist, W. C. Hueper, former Chief of the Environmental Cancer Section, National Cancer Institute, is already convinced of the cancer-causing potential of pesticides. In *Chemical Carcinogenesis and Cancers,* which he wrote with W. D. Conway, he said of chlorinated hydrocarbons such as DDT: "Many of them have distinct hepatotoxic (liver-damaging) properties. Prolonged feeding of rats with DDT and Aramite was followed by the development of hepatomas (liver tumors). While DDT has a minimal tumorigenic effect when given to rats and dogs with the feed in doses highly excessive to those encountered under ordinary exposure conditions, Aramite causes benign and malignant liver tumors in the majority of rats and dogs fed this chemical in the feed at concentrations of 5,000 parts per million. Hepatomas in mice were recently obtained by feeding aldrin and dieldrin, which are chlorinated hydrocarbon pesticides, as well as in rats with methoxychlor.

"The widespread use of aliphatic and aromatic chlorinated hydrocarbons in industry, agriculture and in the home . . . and their distribution as pesticides from airplanes create frequent contacts from various sources with these hepatotoxic agents for members of the general population apart from those associated with the presence of pesticide residues in the foodstuffs of vegetables and animals. Since DDT and other chlorinated pesticides are excreted in the milk, even infants are bound to have some degree of exposure to these agents. They moreover exert a cumulative effect, being stored in various tissue, particularly fat tissue."

The authors pointed out the potential leukemogenic properties, and concluded that the chlorinated hydrocarbons have elicited aplastic anemia and panmyelophthisis (wasting of the bone marrow).

Since the liver-toxic properties of chlorinated hydrocarbons have been pointed out over and over again, it is not surprising that Dr. D. Coda Martin, President of the American Academy of Nutrition, expressed the opinion that the greater number of hepatitis cases, which have increased so alarmingly in recent years, may be caused by DDT sprayed on the leaves of green vegetables. Another physician, Dr. Morton S. Biskind, writing in the *American Journal of Digestive Diseases,* pointed out that in a hospital in which technical chlordane was routinely applied for roach control in the kitchen and the food storeroom, an epidemic of hepatitis had persisted among members of the resident staff for more than three years.[31]

The tremendous increase in cirrhosis of the liver has been equated with the increase in alcohol consumption. But alcohol is a solvent, and while it can damage the liver by itself, it may get a great deal of help from the liver-toxic pesticides.

A University of Miami study of pesticide concentrations in fat tissue at human autopsies turned up some startling results. Dr. William B. Deichmann and J. L. Radomski and Alberto Rey, all of the Department of Pharmacology and the Research and Teaching Center of Toxicology at the University of Miami Medical School, found that twice the amount of DDT, DDE, DDD, and dieldrin were in the fat of patients who died of liver or central nervous disease when compared to that in persons who died accidental deaths.

Pesticide levels were essentially "normal" in other diseases of the liver with the exception of one case of amyloidosis (abnormal deposits of protein in organs or tissues) in which all pesticide levels were extremely high—the highest encountered.

In all groups of cancers of the lung, stomach, rectum, pancreas, prostate, and urinary bladder, the level of pesticides was twice as high as normal. Many were complicated by metastatic liver disease, the Miami researchers said.

In all cases of high blood pressure, concentrations of pesticides ranged from two to three times normal, they pointed out.

The authors said they did not believe the high pesticide levels were caused by ingestion, but by household use of pesticides. They said none of the persons autopsied had any occupational connection with pesticides.

Nevertheless, the autopsies did show that there is a relationship between the amount of pesticides in the body and diseases such as cancer, portal cirrhosis, and high blood pressure.[32]

Dr. Kenneth P. DuBois, Professor of Pharmacology at the University of Chicago, also turned up some disquieting results in his laboratory. He found, according to a report in *Medical World News*, August 23, 1968 (a physicians' journal), that pesticides interfere with the effects of medications, nullifying some and making others more toxic. He found, for instance, that barbiturates are counteracted by DDT.

"Such counteraction has generally been written off by physicians as the patient's natural resistance to the drug," Dr. DuBois commented.

DDT and the other chlorinated hydrocarbons, he found, interfere with the actions of drugs by stimulating the production of liver enzymes that ordinarily detoxify chemicals. To assess the enzyme-stimulating effect of a chlorinated hydrocarbon, the Chicago toxicologist gives laboratory rats barbiturates plus the test insecticide. In one study, the amount of sedative that would normally keep a rat asleep for an hour was effective for only ten minutes.

As for the organophosphates that have been proved by other researchers to inhibit cholinesterase, the nerve-impulse-carrying chemical, Dr. DuBois found that even with the permissible levels set by the Food and Drug Administration for food, the organophosphates do indeed inhibit cholinesterase. They also inhibit the nonspecific aliesterases, enzymes that remove drugs and toxic chemicals from the body.

Dr. DuBois found that compounds affected by these enzymes include the muscle relaxant succinylcholine, local anesthetics like procaine, and the insecticide Malathion.

To evaluate the organophosphates' effects on nonspecific esterases in rats and mice, the toxicologist and his staff added varying amounts of insecticide to commercial rodent diet. Then they periodically did autopsies on the rodents during a ninety-day period, and measured enzyme activity in the livers and blood.

The researchers discovered that after the rodents had ingested the permissible level of insecticide for one week, aliesterase production was inhibited. Two weeks after the researchers substituted an insecticide-free diet, this effect disappeared.

Dr. DuBois did a second set of experiments, and again confirmed his findings. He pointed out that the degree of hazard from the interaction of drugs and pesticides depends upon the margin between the therapeutic dose and the toxic dose of a particular drug. For some drugs, like common headache medications, overexposure would probably do nothing more than prolong the drug's effect, he said. But with drugs like local anesthetics, the pharmacologic and toxic doses are much closer. An ordinary dose of procaine can cause a toxic reaction in a patient with depressed aliesterase activity. The same holds true for the potentially paralyzing succinylcholine.

Organophosphates inhibit the transmission of messages from one nerve to another by interfering with the chemical carrier cholinesterase. That is why it is not surprising that researchers in Melbourne, Australia, reported in the prestigious British medical journal *Lancet* that prolonged exposure to organophosphorus insecticides may cause depressive or schizophrenic reactions, with impairment of memory and of concentration. Measurements of brain waves showed a slowed activity. They concluded: "Organophosphorus insecticides irreversibly inhibit cholinesterase after lengthy exposure."[33]

The organophosphate insecticides commonly used are thimet, phosdrin, chlorthion, Malathion, parathion, schradan, TEPP. The list of other bug killers, fungus killers, mite killers, and weed killers is endless and so are their potential dangers—which, unfortunately are not recognized until too late.

Insecticides in air, water, and earth have traveled hundreds of thousands of miles from the place they were first used, as numerous reports have shown. They may also persist for years in the soil, either in their original form or in broken-down products.

There have not been many measurements of pesticides actually floating in the air. A series of exploratory determinations involving eighteen samples of ambient air collected from four California cities were made in the autumn of 1963. All but two showed measurable amounts of DDT.

According to Dr. Irma West:

"There is no known way in which drift of agricultural chemicals can be entirely eliminated, whether applied by ground or aircraft. Drift of tracers used in air pollution studies have been authenticated as far as 22 miles and further distances could be expected depending on the accuracy of the means for sensing tracer chemicals."[34]

Billy Ray Wilson, Research Professor in Entomology at Rutgers University, New Brunswick, New Jersey, says that if pesticides are sprayed by plane, 50 percent of the chemical is lost to drift. How far the pesticide drifts depends on the size of the droplets and the speed of the wind.[35]

Professor Wilson also maintains that if an accurate estimate of the number of insects in a field could be made, a more accurate amount of pesticide could be used.

The Rutgers entomologist points out that some of the pesticide sprayed from the air finds its way into the waters of lakes. A persistent pesticide introduced into a body of water may concentrate in different parts of the food chain, he says. An instance of such "biological magnification" can be seen in waters that contain 0.02 part per million of a persistent pesticide while producing plankton containing 5 parts per million and fish containing hundreds of thousands of parts per million.

Moreover, he says, fields sprayed with pesticides have plants as well as bugs that absorb the chemical and may again lead to "biological magnification."

Careless spraying of pesticides can lead to unexpected results.

Take the case of the "fire-ant fiasco." A million acres of land were dusted with a concoction of heptachlor designed to remain effective in soil for at least three years. This program continued for over a year before it was revealed that weathering in soil transforms heptachlor into the very persistent and poisonous epoxide. The United States Food and Drug Administration then banned all use of heptachlor on crops.

"It is really frightening," said Lamont C. Cole, Professor of Zoology at Cornell, referring to the "fire-ant fiasco," to realize that ever increasing areas are being treated with new chemicals by persons who do not give a thought for the welfare of—and who are probably unaware of the existence of—the soil organisms on which the very continuation of life depends. . . . I consider it plain irresponsibility to subject the soil biota to hazards like this without intensive preliminary research."[36]

But even research for a number of years cannot protect us from agricultural poisoning. Take the case of Captan, used for more than seventeen years to protect seeds against fungus which rots them. Measured by ordinary toxic effects, Captan is mild. The Food and Drug Administration regulations allow a residue of 100 parts per million to be left on raw agricultural products, a limit set in 1958 after experiments with dogs, cattle, and poultry showed no toxic effects from much higher concentrations. In contrast, the residue limit for an insecticide such as chlordane is three-tenths of a part per million.

But Captan, an FDA researcher discovered, has serious genetic effects on animals. Captan's effect on the reproduction of cells is disastrous, according to the experiments of Dr. Marvin Legator of FDA's Cell Biology Research

Branch. The chemical inhibits the production of DNA, the basic genetic stuff of life, and breaks up cell chromosomes. Dr. Legator, reporting to a symposium in Washington, said that "Captan is only one of a number of chemicals formerly considered safe that are now suspected of causing genetic damage."[37]

HOW MUCH ARE WE BEING PROTECTED AGAINST HARMFUL AGRICULTURAL CHEMICALS?

President Nixon in announcing his 1971 environmental program for controlling pollution said: "Pesticides have caused major ecological problems, highlighted by the decline of several of our bird species. They can be harmful to man himself; in some cases, improper use has resulted in death. Yet pesticides have provided important benefits by protecting man from disease and increasing his ability to produce food and fiber.[38]

"The Department of Agriculture has an extensive research program directed at developing nonchemical means of controlling pests, but to date nonchemical alternatives are available for use in only a very limited number of situations. Because chemical pesticides will be needed for the foreseeable future, the challenge is to use the right pesticides in the right manner and at the right time so that health and environmental damage will be reduced to the absolute minimum.

"At the Federal level, pesticides are currently regulated under the authority of the Federal Insecticide, Fungicide and Rodenticide Act (FIFRA). There are two major problems with the FIFRA. First, it relies almost entirely on labeling to control the use of pesticides and though many pesticide users fail to read the labels and many who do read them ignore the instructions that they contain, this is really the only control we presently have.

"The second major problem with the FIFRA is that the procedures for registration, cancellation, and suspension of pesticides are either inadequate or are very cumbersome. Often several years pass between the time a cancellation order is issued and the time that it becomes effective. There is no way under the present act to stop the sale of a pesticide, the use of which has been suspended. There is no way to keep track of where pesticides are manufactured. Nor is there any control of pesticides which are not shipped across state lines."[88]

Pesticides are among the most deadly chemicals known to man. Knowledge about their long-term effects, their interaction with one another and with other chemicals in the environment, is negligible. Scientific research described in this chapter is casting suspicion that pesticides, herbicides, and fungicides in wide use may damage unborn children, cause cancer, mental illness, and perhaps work havoc on generations to come through damage to genes. In spite of such alarming possibilities, how is a pesticide approved for use in the United States today?

The manufacturers must present their research on the safety of the chemical to the Food and Drug Administration. Pesticide residue tolerances on food are based on the maximum level of pesticide in food that has no discernible effect on test animals, usually rats, during their lifetimes. This "no-effect" level is divided by 100, and sometimes less, to provide a margin of safety. When good agricultural practices allow a lower tolerance, it may be set accordingly. The contribution to the human diet that a particular food makes is also considered in deciding the amount of pesticide residue allowed. Tolerances are therefore reliable to the degree that the test animal and humans react with no more than a hundredfold quantitative difference to the same level of lifelong feeding exposure to the pesticide in question.[89]

Animal-derived foods, which are the major source of pesticides stored in human fat,[40] and fish and game that may contain high levels of pesticides get to our tables without a check anywhere along the line for the chemical.

Dr. John L. Buckley, Chief of the Office of Pesticide Coordination, United States Fish and Wildlife Service, said that of the salmon in Sebago Lake, Maine, more than half of the older fish have stored levels of poisons above that allowed in domestic meat.

"If there is no choice in using pesticides in agriculture, there is also no choice about adequate monitoring of all our foodstuffs or about setting valid tolerances for pesticides in food," Dr. Irma D. West maintains. She said:

"We cannot afford even one mistake which involves all or most of the population."[41]

The American Public Health Association says in its book *Safe Uses of Pesticides,* published in 1968: "What we do not know about pesticides is more important than what we do know. Can pesticides adversely affect the unborn child, the infant or the elderly? What is the effect of insecticides when combined with other so-called inert ingredients? (The term refers only to the effect on the insects.) What is the effect of pesticides on people who are taking prescribed drugs? What is the best way to fight a fire in a warehouse full of pesticides and not have the poisons washed out by water and flushed via storm sewers into nearby streams?"[42]

A corps of 682 FDA inspectors throughout the country does catch pesticide offenses:

● When a terminal manager in Seattle, Washington, observed a pesticide chemical scattered around a shipment of frankfurters, he immediately called the local FDA office. The office called the Washington Department of Agriculture, which sent an inspector out immediately. The residue turned out to be Phorate, an extremely toxic pesticide.[43]

● Seed corn in Laurinburg, North Carolina, was found to contain methoxychlor, dieldrin, and Captan, "pesticide chemicals not in conformity with regulations."[44]

● Carrots from California contained dieldrin in excess of tolerance.[45]

● Chubs, gutted, iced at Monmouth Beach, New Jersey, contained the pesticides DDT, DDE, TDE, and dieldrin "for which no tolerance or exemption on fish has been prescribed."[46]

● Grated cheese in clear glass jars with wraparound paper labels contained high levels of the pesticide BHC. The product was processed in Lowell, Massachusetts.[47]

But what about the food contaminated by pesticides that gets to consumers' tables without detection?

In the spring and summer of 1968, Montana citizens were drinking milk contaminated with the hydrocarbon chlordane. This pesticide is registered only for uses that will not produce residues in milk. It is not supposed to be used in forage for dairy cattle. Yet it was widely used in Montana to kill insects on just such forage.

A staff writer for *Scientist and Citizen,* a publication

supported by eminent but worried scientists, Joseph D. Salvia, interviewed Montana farmers, state officials, and scientists. Tracing the problem back from the discovery of the pesticide in milk to the insecticide practices of the previous season, he found that there was no control of materials sprayed by farmers or the commercial applicators they frequently employ. The labels which read "Do not feed treated forage to dairy animals" were ignored, apparently, and the "education dispensed by presumably expert advisers was contrary to the warnings on the labels.[48]

The chances of discovering high doses of pesticides on food distributed within a state are almost nil. By federal action, according to the FDA Pesticide Residue Program report:

"The FDA's goal was to investigate one percent of the estimated 2,500,000 shipments of raw agricultural commodities in interstate commerce. Twenty-five violative lots were removed from food channels by seizure. Many additional lots bearing illegal residues were destroyed by producers or under state action. Federal actions during the same period included two prosecutions."[49]

Monitoring of pesticides is shared between the Food and Drug Administration and the Department of Agriculture. There have been many instances in which the Department of Agriculture recommended application of a certain chemical, and then the crops were seized by federal Food and Drug inspectors for violations in use of the chemical. Dr. George Kupchik, director of the Environmental Health of the American Public Health Association, said that while division of the duties may be inefficient, he is glad that both the FDA and the Department of Agriculture have the responsibilities. "The FDA favors the consumer and the Department of Agriculture favors the farmer," he said. "If only one department had the power, then either the consumer or the farmer might suffer."[50]

The congressional watchdog agency's General Accounting Office accused the Agriculture Department of not acting aggressively enough to protect the public from "misbranded, adulterated or unregistered" pesticides in a report sent to Congress in September, 1968. The agency reviewed enforcement actions of the pesticides regulation division of the Agricultural Research Service for the fiscal year that ended June 30, 1967. It also included reports on some cases that were followed into 1968.

The study found that the division had not taken action to prosecute violators of the pesticides control law or to obtain recall of misbranded, adulterated, or unregistered products. Moreover, the agency found the division did not even obtain sufficient data to track defective products to different locations in the country.[51]

In 1965, the FDA realized that it was impossible to maintain zero tolerance for pesticides in milk and some other foods. The advocates of pesticides said that improved testing devices made it possible to show even the most infinitesimal traces of pesticides, and for that reason the zero tolerance was dropped. Opponents of pesticides maintained that the food supply in the United States is so saturated with pesticides that nothing today can remain free of them.

WHAT CAN WE DO ABOUT PROTECTION OF FOODSTUFFS FROM PESTICIDES?

President Nixon proposed a Federal Environmental Pesticide Control Act to replace the FIFRA. The administrator of the Environmental Protection Agency would be required, when registering a pesticide, to classify it "for general use," "for restricted use," or "for use by permit only." Pesticides designated for restricted use could be used by trained applicators. Pesticides designated for use by permit only would require the approval of a trained consultant for each application of the pesticide. Applicators and consultants would be licensed by states, and the Federal Government would contract with the states, localities, or nonprofit organizations to provide funds for training applicators and consultants. A pesticide could be registered for both restricted and permit use.

Among other provisions contained in the administration's bill were authority to permit experimental registration of a pesticide; streamlining of the process of appeals from registration, cancellation, and suspension decisions of the Environmental Protection Agency; authority for the administrator to stop the sale of a pesticide if it is in violation of the act; and registration and inspection of establishments manufacturing or processing pesticides. The President also proposed extending the coverage of the act so that a pesticide cannot be used if it has not been federally registered, providing authority for the administrator to regulate the disposal or storage of pesticides and pesticide containers and instituting a requirement that the administrator publish the data on which he has based his decision to register a pesticide.

The President's bill is not perfect. But it is a start. No action has been taken because Congress has been more concerned about the war in Vietnam and war against inflation, and yet the careless or erroneous use of agricultural chemicals could be more devastating to America than either.

We cannot depend upon government regulations and inspectors to protect us. There are just so many inspectors. Even if we had ten times the number of safeguards and inspectors, chances are that some pesticide-contaminated food would still get through.

What is the answer? Better education, of course, but this is not easy. If farmers were to handle pesticides as the government and its extension services recommend, they would all need college degrees in chemistry and the consciences of saints.

The mixing and handling directions for many pesticides are complicated, especially for farm help, many of whom are illiterate. Such safety precautions as going to a dump and burying the containers are time-consuming, to say the least.

Keeping pesticide-contaminated food, even if one is aware of it, from the market, means a great financial loss, and is much to ask of a man who must scratch a living from the land.

President Nixon's recommendations, of course, would alleviate some of these problems.

We could demand that the state and federal governments produce better testing and monitoring of pesticides and food supplies. But if we do not want to share our food with pests, and if we do not want to pay more for our foods, we must compromise. It takes years to prove a pesticide is absolutely safe.

Part of the answer may be found in the suggestion of Dr. George Woodwell, ecologist of the Brookhaven Laboratory, who said on the "Today" show, on television, in 1968, that the great hazards to the future of the human race are still insufficiently recognized. In his opinion, one of the most urgent things that needs to be done is to legislate certain persistent pesticides out of existence and to replace them with pesticides that would effectively control the insect population.

Pyrethrins, for instance, which are produced from a daisylike flower—the pyrethrum—have been effective bug killers for years. They are more expensive than the organochlorine or organophosphate pesticides, but they do not have a residual action. However, they are the only generally available pesticide that can be legally labeled "Nontoxic to Humans and Pets"; and they could be used much more widely.[52]

According to the United States Department of Agriculture, "Pyrethrum is toxic to a wide range of harmful insects, but the toxicity of pyrethrum to man and animals is low. Normally it does not taint materials to which it is applied. These characteristics make pyrethrum an excellent insect deterrent for use around foods and food products."

There are many other nonchemical means of killing insects, from breeding insect-resistant plants to releasing sterile insects that have very effectively cut down the population of their species.

Radiation, light, and harmless-to-man insect viruses have been used with success, and such methods must be developed to replace harmful chemicals.[53]

The fact is that with all our deadly pesticides, we have not destroyed our insect pests. We have just as many bugs today as when we began applying DDT, and the ones we have are rapidly gaining resistance to the chemicals. Furthermore, the Department of Entomology and Economic Zoology at Rutgers reported that "sublethal amounts of certain insecticides will actually increase the rate of reproduction in the female mosquito."[54]

Furthermore, we have succeeded in killing off the natural enemies of these pests. It may be hard to raise sympathy for bats, for instance, but the animals are generally useful and harmless. Insect-eating bats are rapidly being killed off by pesticides. They are among the most susceptible of all animals to the poisons.

Dr. James W. Crow, Professor of Zoology and Genetics at the University of Wisconsin, has listed some serious consequences of the rapid development of resistance:

● The pesticide ceases to be effective . . . larger and larger amounts are needed to control the pest with correspondingly greater upset of the ecological environment and greater risk to man.

● The short useful life of the pesticide means an increasing need for new compounds, so there is less chance for thorough testing and more chances that the compounds will be toxic to man or upsetting to the ecological community.

Dr. Crow suggested that chemical pesticides be applied selectively and used only when really needed. He added that the "hazards to man and to the balance of nature would be made less if a large variety of more selective pesticides could be found. The broader the spectrum of the drug, the greater its risk to other species, including man."[55]

HOW YOU CAN PROTECT YOURSELF FROM IMPROPER USE OF PESTICIDES IN THE HOME[56]

I shudder to think of the many times I sprayed my own kitchen to kill ants, and used insecticide on the rosebushes while my husband watered the lawn, and the children and dog played nearby. So that you won't make the same mistake, or any others with pesticides, here is some advice:

● Read the label and follow the directions. Reread them each time before using. Take the advice seriously. If it says to wear rubber gloves, for instance, do so!

● Use the right pesticide at the right time. Try to pick the least poisonous for the job. Measure accurately. Overdosage seldom kills more insects or diseases.

● Work in a well-ventilated area.

● *Keep pesticides out of reach of the children.* In 1963,

for instance, a year-old infant found lindane tablets for a vaporizer under the sink in her home. She ate four to five tablets. She died at the hospital shortly afterward. In the same city, an eighteen-month-old-boy swallowed half a lindane tablet found on the floor of his family car. He died twelve hours later.

● Do not spray pesticides of any kind anywhere if you are pregnant or around a pregnant woman.

● When spraying inside the home to control flying insects, cover all food and utensils; close windows and doors tightly. After spraying, leave the room, close the door. Do not reenter for half an hour or longer. Aquariums, birds, dogs, cats, and other pets must be removed before you spray.

● Outside, remove or cover food and water containers used by pets. Do not contaminate fish ponds or streams.

● *Do not spray with children nearby.*

● Always keep pesticides in their original containers. Make sure they are tightly closed and plainly labeled. Never put a pesticide in an empty food or drink container of any kind.

● Never smoke, drink, or chew gum while handling pesticides.

● Avoid inhaling sprays, dusts, or vapors.

● Have soap, water, and a towel available. Should you spill concentrated pesticide on yourself, wash immediately.

● When you have finished using a pesticide, wash hands and face thoroughly and remove contaminated clothing before smoking or eating. Workclothes should be laundered before they are used again. Watch your shoes. Sometimes they soak up the pesticide, and should be thrown out.

● Store pesticides and pesticide equipment in a locked cabinet or room that is cool, dry, and well ventilated.

● Never store pesticides with or near food, medicine, or cleaning supplies.

● Some pesticides, such as 2, 4-D or Silvex, may have vapors that can be absorbed by nearby containers.

● Be careful in disposing of containers. Do not burn them. Wrap them in several layers of newspaper and place them in a trash can just before the trash is collected. Until then, lock them up.

● Purchase only what you need for one season.

● If you have a special sensitivity to pesticides, consult a physician and, if necessary, avoid further exposure to the offending chemicals.

● If you experience headache, nausea, or blurred vision, or if you accidentally swallow any pesticide, call a physician immediately. Read the label to him. If necessary, go to a doctor's office or hospital, and take the empty container with you.

● Remember to wash all fruits and vegetables—even those that come prewashed in plastic bags—before eating.

2 NOTES

1. Karl H. Kolmeier, Edwin D. Bayrd, "Familial Leukemia: Report of an Instance and Review of the Literature," *Proceedings of the Staff Meetings of the Mayo Clinic*, Vol. 38, Rochester, Minn., Wednesday, November 20, 1963.
2. *New York Times*, November 27, 1967.
3. *Ibid.*, September 28, 1967.
4. *Ibid.*, January 12, 1968.
5. Irma West, M.D., Berkeley, Calif., "Lindane and Hematologic Reactions," *Archives of Environmental Health*, July, 1967, Vol. 15.
6. *Ibid.*
7. Houghton Mifflin, 1962.
8. Jamie L. Whitten, U.S. Congressman (D., Mississippi), November 6, 1967, National Agricultural Chemical Association, Palm Springs, Calif.
9. *Today's Health*, March 1, 1966, special report on pollution.
10. *1968 Outlook Issue of the Farm Cost Situation*, November 22, 1967, a publication of the Economic Research Service, U.S. Department of Agriculture.
11. C. H. Hoffman and L. S. Henderson, "Protecting Our Food," *U.S. Yearbook of Agriculture*, 1966.
12. *Newark Evening News*, March 24, 1968; Associated Press report, *New York Times*, March 22, 1968, p. 41.
13. New Jersey Community Pesticide Program release, May 12, 1967. Rachel Carson, *Silent Spring*. Karl H. Kolmeier, Edwin D. Bayrd, *Journal of the American Medical Association*, November 20, 1963; *ibid.*, editorial, December 4, 1967, Vol. 202, No. 10, p. 981.

14. *Science News Letter* 89: 136, February 26, 1966.
14a. Barry H. Dvorchik and Thomas H. Maren, M.D., lecture presented to the American Federation of Societies for Experimental Biology, April 15, 1971.
15. Irma West, M.D.
16. R. D. O'Brien and F. Matsumara, National Institutes of Health, General Medical Sciences, HEW report, 1964–65.
17. Irma West, M.D., "Pesticides as Contaminants," *Archives of Environmental Health,* November, 1964, Vol. 9, p. 631.
18. *Ibid.*
19. *Ibid.*
20. *Ibid.*
21. Douglas James, Ph.D., American Institute of Biological Sciences meeting, August 20, 1965.
22. David Metcalf, M.D., New York Academy of Sciences Conference. "The Biological Effects of Pesticides in the Mammalian System," May 2–5, 1967.
23. *Medical World News,* June, 1963.
24. *Science News,* April 10, 1971, Vol. 99.
24a. *New York Times,* January 15, 1969. "DDT Termed Peril to the Sex Organs."
24b. *New York Times,* "Swede Says Intake of DDT by Infants Is Twice Daily Limit," May 6, 1969.
24c. *Medical Tribune,* "Unsafe Doses of Dieldrin Reported in Breast Milk," February 24, 1971.
24d. William B. Deichmann, William E. MacDonald, Dewey A. Cubit, "DDT Tissue Retention: Sudden Rise Induced by the Addition of Aldrin to a Fixed DDT Intake," *Science,* April 16, 1971, Vol. 172, pp. 275–276.
24e. Donald S. Kwalick, M.D., *Journal of the New Jersey Medical Society,* May, 1971.
25. Yank Coble, M.D.; Paul Hildebrandt, D.V.M.; James Davis, Ph.D.; Frank Raasch, M.D.; August Curley, M.S., *Journal of the American Medical Association,* November 6, 1967, Vol. 202, No. 6. Names of patients are fictitious.
26. Frank P. Cleveland, M.D., Cincinnati, "A Summary of Work on Aldrin and Dieldrin Toxicity at the Kettering Laboratory," *Archives of Environmental Health,* 1966.
27. "Evaluation of Some Pesticide Residues in Food," Food and Agriculture Organization of the United States, World Health Organization, 1966.
28. James J. Nora, M.D.; Audrey H. Nora, M.D.; Robert J. Sommerville, M.D.; Reba M. Hills, M.D.; and Dan G. McNamara, M.D., "Maternal Exposure to Potential Teratogens," *JAMA,* December 18, 1967, Vol. 202, No. 12.
28a. Z. W. Polishuk, *et al.,* "Effects of Pregnancy on Storage of Organochlorine Insecticides," *Archives of Environmental Health,* 20: 215–217, February, 1970.
29. Ned D. Bayley, Director of Science and Education, United States Department of Agriculture, before the Senate Commerce Committee on actions and information taken

on the herbicide 2,4,5-T, other phenoxy pesticides, and the dioxins, June 17, 1970.

30. *Progress Against Cancer*, 1966, report by the National Advisory Cancer Council.

31. Booth Mooney, *The Hidden Assassins* (Chicago, Follett, 1966), p. 107.

32. W. B. Deichmann, Ph.D.; J. L. Radomski, Alberto Rey, Department of Pharmacology, University of Miami Medical School, "Retention of Pesticides in Human Adipose Tissue—Preliminary Report," *Industrial Medicine and Surgery*, 37:3, p. 218–219, March, 1968. *New York Times*, March 23, 1968.

33. S. Gerson, M.B., and F. H. Shaw, Ph.D., University of Melbourne and Prince Henry Hospital, Melbourne, Australia, *Lancet*, 1:1371–1374.

34. Irma West, M.D., "Pesticides as Contaminants," *Archives of Environmental Health*, November, 1964, Vol. 9.

35. Billy Ray Wilson, "Spring Scenario," *Re:Search*, Spring, 1969, Vol. 2, No. 2, pp. 8–9.

36. Lamont Cole, Ph.D., "Pesticides: A Hazard to Nature's Equilibrium," *American Journal of Public Health*, January, 1964, Vol. 54, No. 1, pp. 30–31.

37. *Science News Letter*, June 17, 1967, Vol. 91.

38. The President's 1971 Environmental Program, Controlling Pollution, Reform, Renewal for the 70s, the Domestic Council Executive Office of the President, Washington, D.C., 1971.

39. Irma West, M.D., "Pesticides as Contaminants," *Archives of Environmental Health*, November, 1964, Vol. 9.

40. *Ibid.*

41. *Ibid.*

42. *Safe Uses of Pesticides*, American Public Health Association, 1968.

43. *FDA Papers*, July–August, 1967, p. 37.

44. *FDA Papers*, April 1971, p. 38.

45. *Ibid.* p. 42.

46. *Ibid.* p. 42.

47. *FDA Weekly Recall Report*, April 29, 1971–May 5, 1971.

48. Joseph Salvia, *Scientist and Citizen*, October, 1968, St. Louis, Mo.

49. Report: "FDA's Pesticide Residue Program," U.S. Department of Health, Education, and Welfare, Food and Drug Administration, p. 4.

50. Dr. George Kupchik, taped interview, February 26, 1968.

51. *New York Times*, September 17, 1968.

52. Tom Bogaard, chemist and General Manager of Kenya Pyrethrum Co., interview with author, January, 1968.

53. "Protecting Our Food," *Yearbook of Agriculture*, 1966, p. 29; *Medical Tribune*, March 7, 1968; p. 23; *Science News Letter*, 85:377, June 13, 1964.

54. *News Views,* College of Agriculture and Environmental
    Sciences, Rutgers University, New Brunswick, N.J., Janu-
    ary, 1968.
55. University of Wisconsin News and Publication Service,
    February 2, 1966.
56. "Safe Uses of Agricultural and Household Pesticides,"
    U.S. Department of Agriculture Handbook No. 321. Wil-
    liam B. Deichmann, Ph.D., *Industrial Medicine and
    Surgery,* April, 1967. *Safe Uses of Pesticides,* American
    Public Health Association, prepared by the Subcommittee
    on Pesticides, 1968. "A Guide to Safe Pest Control Around
    the Home," New York State College of Agriculture,
    Cornell Miscellaneous Bulletin 74.

3

Your Food—
Plus What?

A doctor's daughter sat down to breakfast one morning and, without drinking a beverage or eating another food, took a spoonful of crisp cornflakes. She put them in her mouth, and swallowed. Within minutes, her uvula, the piece of flesh that hangs down in the back of the throat, swelled. It took great effort for her to move her limbs, and she felt extremely fatigued.

An allergist tested her skin for sensitivity to corn, cow's milk casein, cow's milk lac albumin, but she showed no allergic reaction.

The source of her sudden spell of illness remained a mystery until, one day, she sat down to supper and took a bite of reconstituted dehydrated potatoes. She suffered the same attack of throat swelling, extreme weakness, and fatigue. Again the allergist tested the patient's skin, this time with potato extract. The results were again negative.

Puzzled by the young woman's symptoms, the doctor examined the labels of the box of cornflakes and the box of potatoes. The cornflakes the doctor's daughter had ingested had been treated with BHA (butylated hydroxyanisole) to preserve freshness. The package of potatoes the young woman ate listed BHT (butylated hydroxytoluene, a relative of BHA) as a preservative and antimold agent. All the girl's symptoms disappeared when she stopped eating foods that contained BHA and BHT.[1]

A middle-aged man came to the Headache Clinic at Montefiore Hospital in the Bronx, New York, because he suffered from severe headaches on Thanksgiving, Christmas, and New Year's Day. The possibility of a psychological cause, of course, was strong because on these occasions he always ate and drank a lot and his mother-in-law always came for dinner. The main dish was inevitably turkey.

It took several years to pinpoint the turkey as the indirect cause of the headaches. It turned out that the patient was allergic to penicillin; and, although he lived in New York, he always ate New Jersey turkeys that had been fed on a mash containing penicillin.[2]

From the fertilizer put on the ground to the preservative listed on the container, there are hidden substances in your food. You may not eat the proverbial "peck of dirt," but remember that every year you eat more than three pounds of *intentional* chemical additives with your meals.[3]

The use of additives in foods made in the United States rose from 419 million pounds in 1955 to 661 million pounds in 1965—a gain of 58 percent in ten years. Experts estimate that the use of food additives will climb to 852 million pounds by 1970 and 1.03 billion pounds in 1975.[4]

Most food additives are believed to be safe in the dosages set by the Food and Drug Administration, although many scientists—including former FDA Commissioner James L. Goddard—say we really don't know.[5]

A number of additives, as shown by the two cases mentioned at the beginning of the chapter, may cause an allergic reaction. In fact, Dr. Orval R. Withers, Clinical Professor of Medicine at the University of Kansas Medical School, estimates that about 10 percent of the population has a tendency to become sensitive, in the course of a lifetime, to some food, drug, or antibiotic found in food.[6]

Chemicals are known to cause cancer. The late Dr. Margaret Kelly, pharmacologist at the National Cancer Institute, proved that newborn mice are as sensitive to chemical carcinogens as they are to viral carcinogens.[7]

The fact that a chemical has been determined "safe" is no assurance. One has only to recall the sleeping pill Thalidomide, considered so safe that it was sold without a prescription in more than forty countries. It was widely used because it had no "hangover effect" the next day. More than seven thousand babies in West Germany alone were born with severe defects as a result of the "safe" sleeping pill.

It may be that all our food additives are perfectly safe, but the odds are against it. Dr. Stephen Lockey, a Pennsylvania allergist, has been collecting hundreds of cases of allergic reactions to hidden drugs in foods and medicines. He has compiled a list of 2,764 intentional food additives.[8]

Chemicals are added to food for a variety of reasons:

AS COLORING AGENTS—The natural coloring materials in foods may be intensified, modified, or stabilized by the addition of natural coloring materials, certified food dyes, or derived colors. These chemicals that enhance the appearance of food are considered important for the "esthetic value they add and the psychological effect they have on our food consumption habits."

AS ANTISPOILANTS—Chemicals may be used to help prevent microbiological spoilage and chemical deterioration. There is a growing preference for these food additives.

AS FLAVORING AGENTS—In number—2,112—flavor additives probably exceed all other intentional chemical food additives combined. Of these, 502 are natural and 1,610 synthetic.

AS AGENTS TO IMPROVE FUNCTIONAL PROPERTIES— Chemicals in this classification act as thickening, firming, and maturing agents, or affect the colloidal properties of foods such as jelling, emulsifying, foaming, and suspension. Calcium salts, for example, help the texture of canned tomatoes.

AS PROCESSING AIDS—Sanitizing agents, metal binding compounds, antifoaming agents, chemicals that prevent fermentation, and chemicals that remove extraneous materials are grouped in this classification. Examples are silicones to prevent foam formation in wine fermentation, and citric acid to combine with metals to prevent oxidative rancidity.

AS MOISTURE CONTENT CONTROLS—Chemicals sometimes are used to increase or decrease the moisture content in food products. For instance, glycerin is approved for use in marshmallows as a humectant to retain soft texture. Calcium silicate is frequently added to table salt to prevent caking due to moisture in the air.

AS ACID-ALKALINE CONTROLS—Various acids, alkalis,

and salts may be added to food to establish a desired pH, or acid-alkaline balance. Phosphoric acid in soft drinks and citrate salts in fruit jellies are examples of this chemical control of acid-alkaline balance.

AS PHYSIOLOGIC ACTIVITY CONTROLS—The chemicals in this group are usually added to fresh foods to serve as ripeners or antimetabolic agents. Examples of applications for this purpose are ethylene, used to hasten the ripening of bananas; and maleic hydrazide, used to prevent potatoes from sprouting.

AS NUTRITION SUPPLEMENTS—The use of vitamins, minerals, and amino acids has become widespread. The enrichment of cereal foods alone provides 12 to 23 percent of the daily supply of thiamine, niacin, and iron, and 10 percent of riboflavin recommended for human consumption.

The consumer is unaware that these additives are in the products in the supermarket. Take some random examples:

General Mills' Rice Provence contains among its many ingredients, monoglycerides and diglycerides (emulsifiers), monosodium glutamate (flavor enhancer), butylated hydroxyanisole, butylated hydroxtoluene, propyl gallate (antioxidants), propylene glycol (carrier), and citric acid (sequestering agent.)[9]

General Foods' Whip 'n Chill dessert mix contains sodium caseinate (texturizer), propylene glycol monostereate (emulsifier), acetylated monoglycerides (emulsifier), sodium silico aluminate (anticaking agent) and two preservatives, artificial coloring, artificial and natural flavorings, and a number of other ingredients.

Sara Lee's frozen cherry cream cheesecake contains sorbic acid (preservative) and ascorbic acid (antioxidant). General Foods' Awake, a frozen concentrate for imitation orange juice, contains citric acid (acidulant), gum arabic (thickener), carrageenin (stabilizer), vitamins A, B_1 and C (nutrient supplements), and calcium phosphate (anticaking compound), in addition to flavors and artificial colors.

Without additives, Rice Provence, Whip 'n Chill, and Awake would be utterly impossible. Without additives, Sara Lee's cherry cream cheesecake, because of limited shelf life, might be commercially unfeasible.[10]

There are currently 670 items on the GRAS (Generally Recognized as Safe) list. Shortly after the cyclamates were removed from the GRAS list and banned, the FDA started a new study of the list, asking for comments by interested parties, including the public. So far, no word about changes.

However, in February, 1971, food industry leaders bluntly told the Food and Drug Administration that they reserve the right to add chemicals and other substances to foods without even advising the government. The Food and Drug Law Institute, a liaison between the food industry and Washington officials, said it was a myth that the FDA's list of safe food substances includes all the chemicals added to the American food supply. Representatives of Fritzsche Dodge and Olcott of New York, a manufacturer of food chemicals and flavorings, told the FDA:[10a]

"It should be clear that industry has the right to make its own decisions on the status of any substance, whether or not the FDA has listed it, and that it is under no obligation to request the FDA to express an opinion on unlisted materials."

HIDDEN ANTIBIOTICS

In considering the health effects of additives, it is necessary to go back to the beginning—the unintentional additives added to our food when a pesticide, fertilizer, or growth regulator is used, as discussed in the previous chapters.

The next type of additive to be watched for does not come in a factory-processed human food, but in one producing feed and therapeutics for animals.

T. C. Byerly, speaking before a National Academy of Sciences Conference, in 1967, on "The Use of Human Subjects in Safety Evaluation of Food Chemicals," said:

"As an example of the varied chemicals used in livestock production, I checked the advertisements in *Poultry Meat*, September, 1966. Twelve firms advertised the following kinds of products: coccidiostats, histostats (against histomonads in turkeys), fungistats, prophylactic antibiotics, a tranquilizer stated to prevent aortic rupture, an enzyme preparation to suppress manure odor, a chemical to mask such odor, an antioxidant to protect Vitamin A in feeds. Dips, disinfectants, detergents and sanitizers."[11]

One of the most controversial additives in the category is antibiotics. New data have shown that food animals may retain some antibiotics for as long as forty-seven days and that medication can persist even longer in the kidneys of treated animals. The withdrawal periods established under the Food Additive Amendments of 1958 to ensure that there will be no unsafe residues or no residues of antibiotics have been found impractical.[12]

Why should minute doses of antibiotics in the food we eat bother us? Not only are some persons allergic to the antibiotics used; lavish use of the medications could favor the growth of drug-resistant germs.

As always, we are told we must balance the benefit against the danger.

Since 1950, antibiotics have been widely used at low levels in feeds to stimulate the growth of young chickens and pigs. They are also used to increase egg production. Antibiotics are administered at higher levels in feed or drinking water or by injection in the treatment of various diseases of livestock. In fact, an estimated 2.7 million pounds a year of antibiotic feed additives are used in the United States.[13]

In early experiments, low-level antibiotics usually resulted in at least 10 percent greater weight during early growth and 11 percent greater efficiency in feed utilization. In recent experiments, the response to antibiotics has been less. Untreated growth is as rapid and as efficient as in antibiotic-fed animals.[14]

Many veterinary scientists believe livestock yields would be reduced substantially if antibiotics were withheld, but others contend that, because resistance is engendered, antibiotics in feeds are self-defeating in the long run. C. D. Van Houweling, D.V.M., director of the Bureau of Veterinary Medicine of the Food and Drug Administration, wrote in *FDA Papers*, September, 1967: "It is disquieting to the FDA, among others, that no hard answers are available concerning the use of about 2.7 million pounds of antibiotics annually in medicated feeds. The FDA believes it would be imprudent to persist in such a widespread application without an attempt at systematic understanding of the consequences to the microbial environment of man and animals.

"The seriousness of antibiotic resistance was illustrated several years ago at the very place where antibiotics are so heavily relied on. Hospitals became alarmed over their inability to retain control, with penicillin, over staphylococcal infections. In many cases, overconfidence in penicillin had led to relaxed antiseptic discipline, but it was discovered, in addition, that staphylococcus was manifesting the ability to resist penicillin. The micro-organisms produced an enzyme, penicillinase, which nullified the drug.

"Avoiding unnecessary exposure to antibiotics is prudent not only because of the resistance problem but also because persons and animals can become allergic to antibiotics. Their role in therapy may be precluded if hypersensitivity develops. If used unwittingly in such cases, an antibiotic can elicit a fatal reaction."

Dr. Van Houweling said it is easy to forget that only a sketchy picture exists of the life of the bacteria in the intestines. The normal tract has more than two hundred species, and a balance must exist between them to keep intestinal illness at bay.

The FDA in 1968 proposed revoking the residue tolerance of chlortetracycline and oxytetracycline in fish and poultry. Chlortetracycline is cleared for application to raw poultry, fish, scallops, and shrimps to retard spoilage with tolerances of seven parts per million on uncooked poultry and five parts per million on raw marine products.

Certain antibiotics of the tetracycline class have been found to stain baby teeth, cause sensitivity to light, skin rash, and liver toxicity,[15] as well as interference with kidney function.[16] The residue tolerances for tetracyclines in raw poultry and fish were established on the basis that residue in poultry would be destroyed in cooking and residue in raw seafood would not constitute a danger. However, after years of use, it was discovered this was not true.

Penicillin reactions have produced more than one thousand deaths in the United States. According to Dr. Murray C. Zimmerman of Whittier, California, writing in the *Archives of Dermatology*: "The reactions are increasing steadily in both frequency and severity. Part of this increase may be due to allergies developed by the daily exposure of millions of Americans to penicillin-contaminated dairy products."[17]

Penicillin is widely used to treat mastitis, an inflammation of the udder in cows. At any given time, more than 25 percent of the 26 million cattle in the United States have mastitis. Ironically, although antibiotic treatment has been shown to eliminate one kind of mastitis, in frequent cases it is replaced by another kind of infection.[18]

There is no doubt that penicillin used to treat sick cows can affect the users of dairy products. Two cases of Dr. Zimmerman's prove that.

One, a young man who had an injection of penicillin, developed severe hives, arthritis, arthralgia, and fever. For sixty days, his symptoms were partially suppressed by massive doses of corticotrophin, prednisolone, antihistamines, epinephrine, calcium, and various other medicines. He continued to have hives for nine weeks until he was given Neutrapen, a drug that neutralized penicillin. He cleared up almost completely within forty-eight hours.

However, every time he ate dairy products, he had mild, repeated attacks of his symptoms. At the request of Dr. Zimmerman and another physician, the patient ate dairy products in their presence, and the hives soon followed. As little as eight ounces of milk brought on the attack.

If the patient was given Neutrapen in advance, he was able to eat dairy products without symptoms. One time, the doctors gave the patient an injection of plain salt water which the patient thought was Neutrapen. Upon eating dairy products, the patient again had the hives, confirming the presence of penicillin allergy connected with dairy products.

Another of Dr. Zimmerman's patients, a thirty-five-year-old woman, received a penicillin injection for an industrial injury. She had hives, although she had no previous allergic history. She was given Neutrapen and her hives cleared. She was told to avoid dairy products. A short time later, she had a severe outbreak of hives but denied all penicillin exposure. She finally admitted eating roquefort cheese three hours before the hives began, saying: "I didn't think you could find out so easily that I cheated a little bit."

Government scientists and dairymen, of course, have been aware of the problem of penicillin in dairy products and they have taken steps to cut the amount. Milk is supposed to be thrown out for seventy-two hours after a cow receives penicillin. In 1957, the estimate was that 11 percent of the milk had penicillin in it. In 1966, it was down to 3 percent.[19]

Penicillin is still getting into milk. Researchers from the State University of New York at Buffalo reported the case of a woman who drank a glass of milk and suddenly developed generalized itching, a rash, and a severe headache. She had purchased the milk from a supermarket, which in turn had purchased it from a commercial dairy. The dairy reported that the milk was purchased from numerous local farmers, and a routine check for penicillin content was done about twice a month. When the State University researchers tested the bottle of milk they found approximately 10 units/ml of penicillin.[19a]

In addition, Dr. Frank Rosen, a Maplewood, New Jersey, allergist, and former president of the New Jersey Allergy Foundation, said that as much as three hundred units of antibiotics per gram of tissue have been found in beef. He said that one or two units per gram is enough to cause a reaction and that ten units per gram can cause death in a person "exquisitely sensitive."

The United States does not require antibiotics to be sold to farmers on a prescription basis. The drugs are brought directly from salesmen, and then are administered by the farmer either in the feed or by injection.

The British curbed antibiotics in feed in 1969 after studies in that country found possible hazards to human health. FDA Commissioner Charles C. Edwards appointed an eleven-member committee to study the use of antibiotics in animal feed and medication in May, 1970. In the summer of 1971, the U.S. committee studying antibiotics in animal feed still had not made any recommendations.

About the same time that antibiotics came into use in animal feed, two arsenicals also appeared as feed additives. They are 3-nitro-4-hydroxyphenylarsonic acid and arsanilic acid. When fed without an antibiotic to young, growing animals, either of these compounds affects the growth and efficiency of feed utilization in the same manner as an antibiotic.

Quoting from the statement of Dr. H. R. Bird, of the Department of Poultry Husbandry, University of Wisconsin, in the *American Journal of Clinical Nutrition*, May–June, 1961:

"There is no evidence that organic arsenicals are carcinogenic in any species. But because their use in feeds adds a fraction of a part per million of arsenic to food materials which already contain a trace of arsenic and because a proportion of the fraction of a part per million is presumably inorganic arsenic, and because large quantities of inorganic arsenic applied to the skin for long periods are believed to cause cancer, the Delaney Amendment (1958) prohibits the organic arsenicals in feeds. However, users having prior sanctions can continue to use it by virtue of the 'grandfather clause' [that legal exception exempting violators accustomed to procedures not previously forbidden]. Thus the users with prior sanctions have a sort of monopoly on this alleged method of undermining the public health."[20]

Incidentally, Dr. W. C. Hueper, former Chief of the Environmental Cancer Division, National Cancer Institute, said: "Arsenicals in foods have been responsible for a considerable number of chronic arsenic poisonings among consumers of contaminated foodstuff—especially wine. Recent reports from Germany indicate that vineyard workers who drank wine contaminated with considerable amounts of arsenical pesticides developed, more than ten years after cessation of exposure to these agents, cancer of the skin, lung and liver. These observations on dietary arsenic cancer parallel those previously recorded from Germany and Argentina where [there was] consumption of drinking water polluted with arsenicals. Such neoplastic sequelae have repeatedly been observed after a medicinal consumption of arsenicals. Airborne arsenic in urban atmospheres and arsenical residues in tobacco smoke are among frequent contributors to human exposure."[21]

HIDDEN HORMONES

Ranking in controversy with antibiotics, in animal feed and production, are the female estrogen hormones. The practice of injecting pellets of diethylstilbestrol under the skin of chickens and turkeys was permitted by the FDA from 1947 to 1959. Because it was found that human consumers were getting the residue of the hormone in their poultry, this practice was stopped.

However, permission was given to use diethylstilbestrol as a feed additive for cattle and dienestrol diacetate as a feed additive for chickens. Conflicting reports of whether there is a residue in edible tissues of poultry and beef have been made,[22] as well as conflicting reports about the constant ability of the hormone to affect growth.[23]

From *Agricultural Research,* a publication of the United States Department of Agriculture, March, 1968: "Scientists are attempting to find why growth stimulants markedly improve the performance of yearling steers so they gain weight as rapidly as young calves."

That the growth stimulants save cattle producers money is well established. Research dating back fifteen years has shown repeatedly that gains go up about 15 percent and feed efficiency is improved 12 percent, if steers are properly treated with diethylstilbestrol—stilbestrol for short. Treated cattle also produce meat with more protein and less fat.

"Today, about 80 percent of the fed cattle marketed have been treated with stilbestrol. Despite the clear benefits from treatment, scientists have known very little about how the growth stimulants work."

The maximum allowable dose of DES was doubled in 1970 despite complaints from meat processors that it makes beef watery and stringy. Then, in 1971, Harvard scientists reported in the *New England Journal of Medicine* that the drug caused a rare vaginal cancer in the daughters of seven women who took it during pregnancy.

They recommended greater care in administering such drugs to pregnant women.

Senator William Proxmire (D–Wisconsin) immediately charged that the Food and Drug Administration was breaking the law by allowing residues of a cancer-inducing hormone to reach consumers. Based on Department of Agriculture figures, he said, it was safe to estimate that between 100,000 and 150,000 head of cattle containing residues of the hormone were being slaughtered. He noted that 21 countries and the European Common Market had banned the drug from veterinary use. In addition, Italy and Sweden have prohibited the import of meat from American cattle that have consumed DES and other growth-stimulating hormones.[24]

Dr. D. C. Van Houweling, Director of the FDA's Bureau of Veterinary Medicine, immediately answered Senator Proxmire's charge by saying public health officials have recognized the drug as a cancer agent in laboratory tests. However, the drug, he said, does not show up in beef if farmers withhold it from animals for at least 48 hours before slaughter, as required by the government.

Cancer of the kidneys in male hamsters and castrated female hamsters has been produced with implants of diethylstilbestrol.[25]

Here is what Dr. Hueper has to say on the use of diethylstilbestrol for accelerated fattening of food animals:

"Practical experience has shown that farmers and poultry men do not always insert the estrogen pellets in parts [neck] of animals which usually are discarded, but in parts which are eaten; they implant more than one pellet of 15 mg. of estrogenic chemical, and they sell their animals for human consumption before the safety period of six weeks after implantation has elapsed, while cattle estrogenized by both routes are slaughtered, as a rule, without observing a 60-hour estrogen-free waiting period. It is noteworthy, moreover, that at least one synthetic estrogen, tri-p-anisolcholeroethylene, is stored in the human fat tissue after oral administration and thus appears to be most objectionable from a carcinogenic viewpoint because of its prolonged effect.

"Since members of the general population may have appreciable contacts with exogenous estrogens from other sources (medicines, cosmetics, production of estrogens, handling of estrogenic preparations, preparation and handling of estrogenic feed), such exposures or the possibility of such exposures, especially when they are avoidable and not essential or justified, appear to be objectionable even if it is still controversial whether or not estrogens cause cancers in man.

"It is rather remarkable that biologically potent chemicals which are obtainable for medicinal reasons only on prescription by a licensed physician can be used freely in large quantities by persons without any proper training concerning the potential health hazards associated with the handling and consumption of large quantities of these hormonal substances. Such practices are difficult to control adequately on a nationwide basis in foodstuffs handled in interstate and intrastate commerce by thousands of individual producers in quantities of several million animals."[26]

ADDITIVES IN FOOD PROCESSING

When food processors are criticized about adding chemicals to their products, they always point with pride to the vitamins they put in food. Indeed, the addition to milk of vitamin D has practically wiped out rickets.

Furthermore, it now appears that too much of a good thing can be harmful, which is the phrase defenders of food chemicals always use in defense. Dr. Helen Taussig, former president of the American Heart Association, says there is strong evidence suggesting that prolonged intake of excess vitamin D can lead to heart disease, blood-vessel damage, and even birth defects.

The common feature of these disorders is calcification of the heart valves, sometimes accompanied by narrowing of the aorta, the major artery leading from the heart. Vitamin D has long been known to speed the body's production of calcium, and may sometimes overdo it. Many children who are born with defective hearts—as

opposed to those who develop them later—also suffer
from mental retardation.

In a report of the American College of Physicians, Dr.
Helen Taussig said that early evidence implicating vitamin
D came from England in the early 1950's. The British
were then supplementing each quart of milk with 1,000
units of vitamin D, and were also adding it to cereals,
bread, and flour. Children born of mothers nourished on
this food often had unhealthy calcium deposits and a
peculiar facial expression indicative of subnormal IQ.

When the extra vitamin D was taken out of the foodstuffs and reduced by about half in the milk, the number
of babies born with these defects fell sharply. They are
still born frequently in this country with such defects.

In 1966, National Institutes of Health scientists first
showed that vitamin D crosses the placental barrier.

Scientists at the National Heart Institute and the National Dental Institute are now collaborating on studies of
the effects of vitamin D. The dentists have noticed that
malocclusion is more common today than formerly, and
seems to be increasing in frequency. They believe the
increase may be due to too much vitamin D.[27]

Nutritionists recommend 400 units of vitamin D per day
during pregnancy. But some pregnant women taking vitamin pills and vitamin-enriched milk often consume 2,000
to 3,000 units daily.

Dr. Robert E. Cooke of Johns Hopkins University reported finding fourteen to sixteen cases of children with
serious arterial lesions from overcalcification of the blood
vessels related to excess vitamin D in their mothers' diets.

Dr. Taussig believes that some people inherit a natural
resistance to rickets and that it is these people who are
made sick when extra doses of the vitamin are added to
their diets. Her suspicion is the stronger because the disorder often seems to run in families.

"With our present state of knowledge," Dr. Taussig
warned, "this is a good time to remember that some is
good but more is not necessarily better. People who need
vitamin D should, of course, have it, but there is no need
to make it part of everyone's routine."

If too much vitamin D is bad, too little vitamin A is
unhealthy.

An unexplained and "shocking" incidence of probable
vitamin-A deficiency has been discovered in Canada, one

of the better-fed nations in the world, according to a Canadian nutrition scientist in a report made at the Western Hemisphere Nutrition Congress in Puerto Rico, in August, 1968. The evidence was gathered in autopsy examinations of livers, said T. Keith Murray, Ph.D., Chief of the Nutrition Division, Canadian Food and Drug Directorate. Liver is a storage organ for vitamin A; and a liver seriously depleted of vitamin-A stores is taken as a sign of vitamin-A deficiency.

Postmortem examination of more than a hundred persons ten years of age or older revealed that more than 30 percent of them had no more vitamin A in liver storage than is customarily found in newborn infants.

"Clearly a high proportion of Canadians were in unsatisfactory vitamin-A status at the time of death," he said.

"It is hard to believe that this is entirely a dietary problem. The 'apparent' per capita consumption of vitamin A in Canada is 6800 International Units per day, well above minimum requirements. Even allowing for wastage, cooking losses and uneven distribution it does not seem likely that so many of our population do not get enough vitamin A to maintain their reserves.

"No clear pattern relating low stores to a particular disease has emerged, however, and we suspect that some environmental factor—drugs, pesticides, food additives, etc.—may be reducing the utilization or increasing the metabolism of vitamin A."[28]

BABY FOODS

If any food should be pure and without potentially harmful additives, it should be food that is intended for infants. The World Health Organization has cautioned that baby foods "require separate consideration from all other foods as regards the use of food additives and toxicological risks."

The WHO committee pointed out that infants don't have fully developed systems and cannot eliminate poison-

from the body as adults do. In addition, if a baby eats the same concentration of an additive in his food that an adult does, it will result in a higher concentration in his body.

One additive commonly found today in baby foods is sodium nitrite, used to preserve and produce the pink color in cured meats. Nitrites reduce the blood's ability to carry oxygen from the lungs to the body tissues. Nitrates, which are unintentional additives to baby foods (although not to adult foods), become nitrites when converted by intestinal bacteria.

An editorial in *The Lancet*, highly respected British medical journal of May 18, 1968, warned of a possibility that under certain circumstances foods preserved with nitrites may combine with amines in the intestinal tract during the process of digestion and with nitrosamines and N-nitroso compounds, most of which have been proved to be strong carcinogens.

Dr. Samuel Epstein, an environmental pathologist and toxicologist at Harvard and the Children's Cancer Research Foundation of Boston, and Dr. William Lijinsky, of the Eppley Institute for Research in Cancer at the University of Nebraska College of Medicine, have said repeatedly since 1968 that their experimental evidence shows when nitrites combine with natural stomach chemicals, amines, potent cancer-causing compounds, are produced.[28a] In 1970, two other researchers, Dr. Melvin Greenblatt and Dr. Sidney Mirvish, of the University of Nebraska, reported lung cancers were produced in 65 to 90 percent of mice fed large amounts of nitrite.[28b] On May 17, 1971, an article in the *Journal of the American Medical Association* entitled, "Common Compounds Linked to Cancer," described how physicians around the world have linked nitrites and nitrates to cancer. However, the Food and Drug Administration has done nothing to take nitrites and nitrates out of baby food. The compounds are also widely used in curing adult foods such as processed meats and fish.

Nitrates have another unhealthy effect. They cause methemoglobinemia, a condition that cuts off the supply of oxygen to the brain. When pregnant women ingest nitrites, there is a possibility that they can cut off the supply of oxygen in the blood to the fetus, according to research reported in New York.[28c]

A highly respected scientist, Dr. Barry Commoner, Director of the Center for the Biology of the Natural Systems, University of Washington, St. Louis, Missouri, said: "Sufficiently concentrated dietary nitrate can lead to respiratory failure, and even death." He cautioned that infants are particularly susceptible to this hazard because their intestinal bacteria are more likely to include types that convert nitrate to nitrite.

Dr. Commoner, who sounded some of the earliest warnings about the danger from nuclear fallout, said a study of commercial baby food conducted by the Missouri Agricultural Experiment Station showed that certain vegetables—beets and spinach—contained as much as 0.8 percent nitrate.

"At this rate," the biologist calculates, "an infant fed a two-ounce jar of baby food would receive about 40 milligrams of nitrogen as nitrate." But, he added, "public health officials recommend that infants take in [in] their food and water no more than 12 milligrams of nitrate nitrogen daily."

Research by European scientists, Dr. Commoner said, places the blame for the excess of nitrates in the food directly on chemical fertilizers. Spinach grown without chemical fertilizer contains less nitrates than spinach grown with it.[29]

Dr. Commoner did not mention the sodium nitrite deliberately added to baby food.

In one brand of meat for infants, the author found ascorbic acid listed as well as sodium nitrite. *The Merck Index*, an encyclopedia of chemicals and drugs, states in the Eighth Edition (1968) that ascorbic acid should not be formulated with sodium nitrite. The reference is based on Russian scientific work done in 1952 which showed that ascorbic acid makes sodium nitrite more potent. (In some packaged meat for adults, the author has seen ascorbic acid, sodium nitrite, and sodium nitrate listed among the numerous additives.)

Although this author is not aware of any current studies going on, the possibility that nitrates and nitrites may be the indirect cause of "crib deaths" certainly should be investigated. More than 20,000 crib deaths occur annually in the United States, when parents put an apparently healthy child to bed and then find the baby dead some time later. Theories of causes range from undetected heart disease or self-inflicted whiplash injury to infanticide, smothering, and allergy. No one, as yet, has proved a cause.

Too much of another common substance is evidently blithely being added to baby foods. Here is a direct quote from Philip L. White, Secretary of the Council on Foods and Nutrition of the American Medical Association, from the AMA publication *Today's Health,* July, 1968:

"It has been suspected for some time that the sodium content of baby foods is higher than necessary. Salt is added to most baby foods to enhance their flavor. Whether this is done for the benefit of the child or the mother is debatable. Most likely, it is for the mother's benefit, since the baby's taste perception is slow to develop.

"Baby-food manufacturers design their products to be attractive to mothers. If a food item is not acceptable to the mother, she will not use it or will over-react to apparent rejection by the baby and will not buy it again. Therefore, salt is added in fairly heavy amounts to most baby foods. The taste and color are developed to attract the mother while the consistency is developed to accommodate the baby.

"A study of the sodium intake of infants found that infants two months of age received essentially all of this sodium from milk. Infants from five to seven months received about 50 percent of their sodium from milk and 40 percent from vegetables and meat, and eggs (chiefly in prepared baby foods). The average daily sodium intake was just under one gram. Youngsters about one year old derived 60 percent of their sodium from vegetables, meat and eggs and 30 percent from milk. At this age, the usual daily intake of sodium was 1.4 grams. This would seem to be a large amount of sodium.

"Evidence is beginning to accumulate suggesting a relationship between high sodium intake and early development of hypertension (high blood pressure). Studies have demonstrated that high intakes of sodium will hasten the onset of hypertension in strains of animals susceptible to the disease.

"Pediatricians, while not unconcerned about the rather large amount of sodium in the diets of infants, feel that until more convincing evidence of hazard is available, there is little reason to criticize current manufacturing procedures. The infant can excrete the sodium not required in metabolism without apparent difficulty so long as his kidney function is adequate and his water intake is ample. On the other hand, there is little to be gained by giving the infant sodium far in excess of his requirements. This is particularly pertinent since salt is added for flavor only."

To quiet mothers' fears about too much salt, Gerber Baby Foods issued a pamphlet in 1971, which said: "The 1969 White House Conference Panel on Pregnant and Nursing Women and Infants found 'no scientific basis upon which to recommend prohibiting the practice of adding table salt in infant foods at levels now in practice.'"

It continued: "Since that time, a special National Academy of Sciences committee was authorized by the Food and Drug Administration to study the use of salt in infant food. As previously agreed, the baby food industry will follow the recommendations of this committee."

Gerber and the other baby food manufacturers did not wait for government action to remove another additive in baby food, monosodium glutamate, in 1969.

John W. Olney and Lawrence G. Sharpe, of the Department of Psychiatry, Washington University School of Medicine, St. Louis, Missouri, reported in *Science*, October 17, 1969, that infant rhesus monkeys given monosodium glutamate suffered brain damage. Charles U. Lowe of the National Institute of Child Health was quick to dispute the finding by pointing out that the monosodium glutamate was injected, not eaten, and that an infant would have to consume 20 jars of baby food to obtain an equal amount of MSG.[29a]

Olney and Sharpe's work was then confirmed by researchers at Harvard, and the FDA formally asked for a study of MSG by the Food and Nutrition Board of the National Academy of Sciences–National Research Council.[29b]

The baby food manufacturers removed MSG without waiting for a verdict. But why did they put it in baby food in the first place?

Still another additive to which children may be exposed in ice cream, jellies, chocolate drinks, icings, and certain candies is carboxymethyl cellulose. Made from a cotton by-product, it is added as stabilizer. It was in every one of the highly advertised mixes for children to put in their milk that the author found on the supermarket shelves. Carboxymethyl cellulose has been shown to cause cancer in animals.[30]

Fredrick J. Stare, M.D., Professor of Nutrition and head of the Department of Nutrition at Harvard School of Public Health, and Dr. Charles A. Janeway, Professor of Pediatrics at Harvard Medical School, wrote an indignant article that appeared in the prestigious *New England Journal of Medicine*, on September 7, 1967, entitled: "Are All Baby Foods Special Dietary Foods?" They referred to the pending revision of regulations for food of special use by the Food and Drug Administration.

"We are disturbed . . . by the interpretation that may be given those responsible for the enforcement to the definition of Special Dietary Foods which would place ordinary baby foods in this classification.

"There are few cases in which infant tolerances for certain food additives—for example, calcium silicate—are considered to be lower than adult tolerance. These differences do not appear to be valid reasons for classifying all baby foods as special dietary foods, particularly since these situations have been handled quite satisfactorily by the setting of specific tolerances for their use in baby foods under the food additive regulations.

"Those whom we have consulted in the 'baby food' industry contend that the reclassification of baby foods as special dietary-purpose foods would work a hardship on current business practices and serve no advantage to the consumer. Reclassification would probably make necessary certain changes in labeling (for example, it would be necessary to note the source of all ingredients and to specify all flavoring ingredients by their generic names). This would add more copy to labels that are already too small to be read with ease."

To that last statement, as a mother of allergic children, I am distressed that a nutritionist of international renown and a respected pediatrician could make that statement. Certainly, everything that is put into baby-food jars should be described on the label. If necessary, make a bigger label or even a bigger jar.

As for baby foods being under a special dietary protection, a government official answered the article in the *New England Journal* for the following month by pointing out that strained and chopped fruits and vegetables marketed as foods for infants are, in fact, classified as foods for special dietary use and have been so since 1941.

I don't know whether many other doctors agree with the two Harvard physicians, but few mothers would like to see baby foods exempt from the special attention given dietary foods—imperfect as it is—by the Food and Drug Administration.

The question FDA officials and baby food manufacturers must answer is whether the additives they are allowing to be added to baby food are really *necessary* and whether they are, in fact, potentially harmful.

ARTIFICIAL SWEETENERS

When the United States Secretary of Health, Education and Welfare restricted the use of the artificial sweetener, cyclamate, October 18, 1969, more than 175 million Americans, including young children, were ingesting the chemical. The rise in the use of cyclamate, which was intended originally only for diabetics and others on sugar-restricted, medically supervised diets, was from 0.25 million pounds in 1955 to more than 17 million pounds in 1969. Sixty-nine percent of the artificial sweetener was used in beverages, 19 percent in table sweeteners, 6 percent in foods, 4 percent in non-food items, and 2 percent was exported.[31]

Cyclamates were first marketed in the early 1950s, for diet foods. In 1958, they were placed on the GRAS list so that no restrictions were imposed on their use.

As early as 1962, scientists began to have doubts about the unrestrained intake of cyclamates. In the fall of 1964, the *Medical Letter*, published by a group of prominent American physicians, asserted that excessive use of these sweeteners is against the public interest.[32]

In 1965, the Wisconsin Alumni Research Foundation reported after a diet of 5 percent calcium cyclamate for nine months and 95 percent normal nutrition, laboratory rats grew 12 percent less than a control group not receiving the sweetener. When the diet was changed to include 10 percent cyclamate, the rats suffered a growth impairment of 50 percent.

In 1966, FDA researchers studied the sweeteners after Japanese scientists said that birth defects associated with artificial sweeteners were found in experimental animals. Tests to verify this finding, which are expected to take years, are being done at Albany Medical College.

The National Academy of Sciences announced that the new study is "appropriate in the light of the new toxicological information on the increased use of artificial sweeteners."

Research at Albany had already shown that cyclamates caused symptoms imitating hyperthyroidism, and result in patients being treated for the hormone disorder erroneously. Furthermore, the Albany researchers found that some people convert cyclamates into cyclohexylamine, a substance that raises blood pressure.[33]

On June 5, 1969, scientists at the University of Wisconsin reported to Abbott Laboratories, manufacturer of cyclamates and sponsor of the research, that a significant incidence of bladder tumors had been found in white Swiss mice in two experiments in which pellets of cholesterol and cyclamates were implanted into the lumen of the urinary bladder. Representatives of Abbott Laboratories discussed the matter with representatives of the National Cancer Institute and the Food and Drug Administration and they came to the conclusion that the cancer-causing properties of the pellet did not convict cyclamates that were given orally, but that further studies should be performed.[33a]

On October 8, 1969, an Abbott scientist was notified that there appeared to be bladder lesions in rats fed a 10:1 mixture of cyclamate sodium/saccharin sodium over a two-year period. This was based on their contract-supported experiments at the Food and Drug Research Laboratories, Long Island, New York. During this study, many of these rats were shown to be able to convert cyclamate to cyclohexylamine (CHA).[34]

On October 13, representatives of Abbott Laboratories met with FDA researchers to review the study of cyclamate sodium/saccharin sodium mixture. Of 240 rats, seven males and one female showed bladder tumors rarely seen in rats.[34a]

On October 18, HEW Secretary Robert H. Finch ordered the artificial sweetener cyclamate removed from the Generally Recognized as Safe list.[34b]

Since the ban, further confirming evidence of the cyclamates' cancer-inducing properties have developed, and new evidence showing the chemical's ability to break chromosomes and cause deformities of chick embryos has been discovered.[34c]

After the ban, the FDA banished to the boondocks Dr. Howard L. Richardson, its chief pathologist, who had made public a 1950 agency study on the questionable safety of cyclamate, the artificial sweetener. The 1950 experiment showed the same cancer dangers that led to its ban by the government in 1969.[34d]

What about saccharin, the artificial sweetener that has been substituted for cyclamate in most of the artificially sweetened products? Discovered in 1879, saccharin is pound for pound about three hundred times as sweet as natural sugar. Cyclamates have about thirty times the sweetness of sugar, but were preferred to the older saccharin because they did not have saccharin's characteristic bitter aftertaste.

Saccharin was used along with cyclamates in the experiments that led to the ban on cyclamates. The same researcher who conducted the experiments, Dr. George T. Bryan, of the University of Wisconsin, has charged that saccharin causes more dangerous tumors in mice than cyclamates, and should be banned.[34e]

The FDA has asked that the National Academy of Sciences–National Research Council review material available concerning the safety of saccharin. A preliminary report by NAS–NRC said that there is no immediate danger and that saccharin should be studied further. The NAS–NRC had followed a similar path in judging cyclamates. The National Cancer Institute is also conducting a study with saccharin and its report on the subject is due in late 1971.[34f]

In the meantime, millions and millions of people are being exposed to possible cancer-causing artificial sweeteners as they have been for more than twenty years, and the incidence of bladder cancer has doubled in men in the past twenty years.[34g]

A substance being considered as a replacement for the cyclamates is hesperidin, one of the bioflavonoids. United

States Department of Agriculture researchers who developed the process said it appears preferable to currently marketed artificial sweeteners because it is a natural ingredient in all citrus fruits.[35]

However, since the citric sweetener was synthesized by Department of Agriculture researchers, it is in the public domain and no company can get an exclusive patent on its use. As a result, no company, thus far, is interested in producing it.[35a]

The U.S. Army is also experimenting with a small red berry from "Miracle Fruit." Ghanaians have been using it for centuries. It produces a sugar taste and has no calories, but the sweet aftertaste lasts for hours.

CHEMICALS IN COMBINATION

During 1965 and 1966, there appeared to be an "epidemic" among apparently healthy Canadians, Americans, and Belgian men in their mid-forties who suddenly dropped dead. The only thing they had in common was a love of drinking beer. Autopsies showed that the beer drinkers had a peculiar destruction of their heart muscle.

Detective work by physicians and government scientists turned up the fact that a month before the beer drinkers began to die, some beer manufacturers began to add cobalt, a metal, to their product to improve the quality of the foam.

University of Michigan researchers sought to determine whether cobalt additives could, indeed, cause heart disease. James L. Hall and Dr. Edward B. Smith, who conducted these studies in animals, presented their findings to the annual meeting of the American Association of Pathologists and Bacteriologists in Chicago in 1968.

When rabbits were given cobalt in beer, they said, the animals' hearts took on a "moth-eaten" appearance, "with focal areas of noninflammatory degeneration marked by interstitial edema, separation of fibers, vacuolation and fibrillar degneration. Loss of cross-striations was also a prominent finding."

These lesions were the same as those described in the hearts of the unlucky beer drinkers.

When the cobalt was given without the beer, there were no such changes in the animals' hearts, leading the researchers to conclude that the combination of cobalt and alcohol caused the damage.[36]

One of the dangers that are becoming more and more a matter of concern among eminent scientists is the effects of chemicals that singly may be harmless but in combination may be extremely toxic.

It is known that a sensitive balance exists among the microbes that normally reside in the human gastrointestinal tract. Dr. R. E. Eckhardt, of Esso Research, a physician, questions whether a hangover is really due to the fact that the alcohol consumed produces cerebral edema or whether it is a result of toxins liberated from microorganisms that become more active because they may not be as sensitive to the alcohol as other organisms. He questions what effect the ingestion of perhaps unknown toxins in other foods we eat may have on this sensitive balance within our gastrointestinal tract.[37]

This chemical interaction in food has been dramatically demonstrated in the more than thirty-one court cases of the "depressed cheese eaters." After taking Parnate, a medicine for combating mental depression, and then eating cheese, these people suffered severe high blood pressure that resulted in symptoms ranging from convulsions to strokes. Parnate inhibits the brain chemical monoamine oxidase (MAO). Cheese and other foods, such as beers, wines, yogurt, and beans, contain the natural chemical tyramine. Together, Parnate and tyramine combine to explode the blood pressure.

Another result of the strange effects of chemical combinations is now being called the "Chinese food syndrome."

A Cantonese doctor, Robert Ho Mas Kwok, enjoyed Chinese food in the Far East and Europe. But when he came to the United States in 1960, he had an odd reaction. He reported in the *New England Journal of Medicine* that every time he eats in a Chinese-American restaurant he develops a strange numbness of the neck, back, and arms that lasts for about two hours, with no hangover effect.

In the following issue of the *New England Journal,* ten other letters, including seven from physicians, agreed that some people were indeed affected by a "Chinese food syndrome." Among the symptoms the writers described were headaches, heavy sweating, weakness, and numbness.

Richard Lyons, a science writer for the *New York Times,* interviewed Chinese restaurant owners in New York, and asked them about the phenomena described in the medical journal.

One restaurant owner commented, "The only headaches I get are from running this place and paying taxes." Only one owner admitted that he had heard of three or four cases of a strange reaction to Chinese food among thousands.[88]

Dr. Kwok, who is a senior investigator with the National Bio-medical Research Foundation, said he had discussed the "Chinese food syndrome" with some Chinese friends who had also been affected by it. They speculated that monosodium glutamate seasoning or soy sauce or cooking wine might be the cause. Another possibility, he said, is that the high salt content of Chinese food may produce a temporary excess of sodium in the blood that reduces the potassium levels of the body. This, in turn, can cause a numbness of the muscles, generalized weakness and throbbing. The condition is termed "hypokalemia."

Another letter writer to the *New England Journal,* Dr. Herbert H. Schaumburg, of the Bronx, New York, agreed that an excess of sodium "sounds reasonable." He said that three times he had experienced a tightening of the face and temple muscles, numbness, weeping, and even fainting after eating "in my favorite Chinese restaurant." (One wonders why he doesn't give up eating Chinese food, especially in that restaurant.)

MUTAGENS

The effects of chemicals in our food singly and in combination are far from a laughing matter. The effects

cursed children

may irreparably damage human beings for generations to come

Dr. James Crow, of the University of Wisconsin, Chief of the National Institutes of Health's Genetic Study Section, said that exposure of large numbers of people to a chemical mutagen could cause a "genetic emergency" or even a "genetic disaster."[39]

Unlike teratogens, which damage an already conceived infant, and afflict only a single generation, mutations occur in the heart of the heredity cells. The damage could lie dormant for several generations, then affect many generations thereafter. The first warning of such an illness, Dr. Crow said, could be an increase in the number and incidence of severity of genetic illness; a marked change in the birth ratio of men to women, or an almost inperceptible loss of vigor and vitality, progressing from generation to generation.[40]

Recent United States and Canadian studies of human conception indicate that at least 25 percent of human conceptions end in intrauterine deaths. Among these spontaneous abortions, chromosomal anomalies can be found in 30 percent, and in 8 percent of all conceptions.[41]

Mongolism and other chromosomal abnormalities are found in at least 1 percent of live births. Another 5 percent of live-born infants have genetic diseases of yet undetermined origin, including diabetes and cystic fibrosis— thus, about one conception out of every six conceptions results in a dead or defective child due to genetic causes.[42] About 50 percent of the dead or defective babies can be shown to carry a "bad" gene the origin of which is unknown. Another 5 percent have known causes such as X ray or viruses. What causes the balance of defective babies is unknown, but researchers suspect that much of this fetal wastage is due to new mutagens. No one knows what, if any, part of these mutations is caused by chemicals.

However, Dr. Crow and a number of other scientists interested in genetics held a meeting at Jackson Laboratories, Bar Harbor, Maine, in 1967, to discuss chemical mutagens. At the request of the National Institutes of Health, the report was held back from publication, reportedly because the federal government wanted to have some corrective measures ready before it was released. Here are some of the points the scientists made:

● A number of chemicals—some with widespread use—are known to induce genetic damage in various organisms.

● Identity of the genetic material in all organisms implies that a chemical that is mutagenic to one species is likely to be so in others, and must be viewed with suspicion.

● Some components are highly mutagenic in experimental organisms in concentrations that are not toxic and that have no direct effect on fertility, so that no natural barrier prevents mutations from being passed on to future generations.

● Perhaps most insidious are compounds that induce gene mutations without chromosome breakage, and thus cannot be detected microscopically.

The chemicals that most concerned the Bar Harbor conferees as potential mutagens included chemosterilants and other pesticides, food additives, cosmetics, air pollutants, herbicides, known carcinogens, industrial chemicals, drugs, vaccines, and contraceptives.

For a long time it has been known that caffeine breaks chromosomes in lower animals and mammalian cells when given in large amounts. Americans over the age of fourteen years consume caffeine in coffee, tea, cola, and drugs at an average rate of 500 milligrams a day.[43] Caffeine may be residually present in the body and appears to retard the natural repair mechanism for broken chromosomes. Its destructive effect may be synergistic with other chromosome breakers, such as X rays.

A New York University chemist reported in 1970 that a common food additive, the preservative sodium bisulfate, can affect the chemicals that comprise hereditary material.[43a]

Captan, as mentioned in the first chapter of this book, is one of the most popular and supposedly least toxic fungicides. For more than seventeen years, it has been used freely on gardens and in orchards. Yet FDA researchers found that it inhibits cell division and is extremely teratogenic in chickens. It pops the eyes out of one kind of fish. At last report, the FDA was studying the matter.

Ironically, information on the mutagenicity of drugs and food additives had not in the past been part of the requirements for approval by the FDA.

A Dow Chemical Company biological researcher said

that the cost of testing two or three highly suspicious chemicals for mutagenicity could reach $500,000 per chemical, even in the animal stage and with the use of computers.[44]

CANCER

If chemicals can cause genetic damage, they can also cause cancer, because both have to do with the abnormal growth of the cell.

Dr. Philippe Shubik, formerly Professor and Director of the Division of Oncology at the Chicago Medical School Institute for Medical Research, and now Director of the Eppley Institute for Cancer Research, University of Nebraska, said: "Within the past decade, particularly, it has become apparent that chemical carcinogenic factors may well be responsible for a large number of cancers in man. An expert committee for the World Health Organization, writing on cancer, concluded that at least 50 percent of all cancers in human beings could be considered to be caused by extrinsic environmental agents.

"It is our belief that many, if not all the majority of cancers in human beings will eventually be shown to be caused by defined agents, and that many of these will be preventable."[45]

Dr. Shubik's colleague in the Division of Oncology, Dr. William Lijinsky, said that the "idea that some cancer in human beings is caused by exposure to chemicals is difficult to deny."[46]

Seventeen years ago, the cancer incidence rate was a little over 250 per 100,000 per year. Projecting the rates into 1985, the insurance companies expect the ratio to reach about 375 per 100,000.[47]

Dr. Morton L. Levin, Director and Assistant Commissioner for Medical Services, State Department of Health, Albany, New York, when testifying before a congressional committee studying food additives in the late 1950's, said: "We must remember that cancer in the human population today, particularly in the United States, will affect one out of four individuals *at some time during life*. We cannot rule out the possibility that substances which we do not today suspect may actually be causative of cancer.[48]

After the hearings, which were conducted by Congressman James Delaney, the Food Additives Amendment, Public Law 85-929, was passed on December 6, 1958. The law specifically states that no additive may be permitted in any amount if the tests show that it produces cancer when fed to man or animals or by other appropriate tests.[49]

This part of the law has been severely attacked by food and chemical manufacturers and even by the Nutrition Council of the American Medical Association. The AMA Council on Nutrition "urged either repeal or revision of certain clauses in the Food Additives Amendments and the 1960 Color Additives Amendments. Both clauses prohibit the setting of tolerances for the use of cancer-causing agents in food.

"These clauses could prohibit the addition of certain essential nutrients to foods if 'any amount' of the substance was shown to cause cancer. Technically, these clauses contribute nothing to the safe use of food additives since any hazardous use of an additive is already prohibited in the general provision of the food additives amendment."[50]

Fortunately, the FDA has withstood the attacks against the "no cancer" clause. The Secretary of Health, in defending the clause, said: "No one really knows how to set a safe tolerance for substances in given foods when those substances are known to cause cancer when added to the diet of animals."

Eminent scientists, including those on the National Academy of Sciences Food Protection Committee, state that no one, at this time, can tell how much or how little of a carcinogen would be required to produce cancer in a human being or how long it would take the cancer to develop.[51]

The National Academy of Sciences Food Protection Committee, while agreeing that low doses of a known carcinogen have been applied in small amounts to mouse skin without cancer developing, stated:

"The possibility exists that doses at 'no effect' levels do in fact exert carcinogenic effects but that the effects are too weak to detect with the numbers of animals feasible for routine testing. Such a possible effect, though extremely weak, might become evident in a large population such as, for example, the population potentially exposed to food additives in our culture. There is also the possibility of synergistic effects among substances present in the diet and of an individual susceptibility to carcinogens, although little is known about these factors.[52]

It would be wonderful if the Delaney Amendment had protected our food against even known carcinogens. Unfortunately, potentially cancer-causing chemicals are still being added to our food.

Dr. William Hueper, that voice crying in the wilderness, said: "Several thousand different chemicals and chemical mixtures used as additives and pesticides for a great variety of purposes are now incorporated into foods. Many of these have not been adequately tested for carcinogenic properties. There is practically no feasible escape route left for the captive population to avoid continued contact with unintentional food additives."

He said the evidence at hand indicates that several food and cosmetic dyes are probably human carcinogens and that others are potential carcinogens. He added that many additional ones have not been adequately tested. Here are some of the food additives and their carcinogenic properties that Dr. Hueper lists in his book *Chemical Carcinogenesis and Cancers*, written in collaboration with W. D. Conway:

"EMULSIFYING AGENTS AND SHORTENINGS—An emulsifying agent prepared from a vegetable oil by the application of heat and oxygen, containing highly oxidized and highly polymerized aliphatic compounds and used for the processing of vegetables and animal fats used in the manufacture of margarine, was shown to produce sarcomas in rats at the site of subcutaneous introduction. Distinct caution, therefore, is indicated in the incorporation of highly oxidized oils such as those probably formed in the deep-fat frying process into products of human consumption. Liver tumors were produced in 40 percent of mice fed a commonly used shortening and in 65 percent of mice given orally a specific solvent fraction of this foodstuff.

"One experiment using polyoxyethylene stereate proposed for use in bread and rolls produced, when fed at 26 percent level in the diet, not only bladder stones in 25 out of 150 rats, but in 13, bladder tumors.

"SYNTHETIC MUCILAGES, THICKENERS AND STABILIZERS—The recent demonstration of various cancerous responses in rats by the parental introduction of water soluble, highly polymerized compounds such as polyvinyl pyrrolidone, carboxymethol cellulose and dextran should provide an indication for an intensive and competent investigation into potential similar responses by some of the similar synthetic polyglucoses. Since such products have no nutrient value, it would be wise to eliminate them for the time being from the list of permitted food additives. [They are still on the list.]

"FLAVORING AGENTS—No comprehensive information is available concerning potential carcinogenic properties of the approximately [2,112] flavoring agents employed. Lemon oil, however, has recently been reported to be a co-carcinogen. Safrole, which served for many years as a flavoring agent of soft drinks, especially root beer, was shown to cause cancer of the liver when fed to rats kept on a normal diet.

"SURFACANTS—Employed in food as antifoaming agents, emulsifiers and dispersants or may be introduced unintentionally into them as a residue of detergents or used for cleaning cooking utensils and dinner ware. In experiments on animals, it has been shown that some chemicals of this type exert carcinogenic or weakly carcinogenic effects on the action of known polycyclic aromatic hydrocarbons given by mouth and facilitate the penetration of these agents through the mucosa of the alimentary tract and skin. Candies, soft drinks, dill pickles, vitamins, ice cream, cream whip, cakes, bread and rolls."

Dr. R. E. Eckardt, of Esso Research, said in 1966 that nitrosamines produced a variety of tumors at different sites. Some of them produced tumors after a single dose in animals. He said it is believed that certain nitrosamines may occur in tobacco smoke but that more recently they have been demonstrated in "herring meal preserved with nitrite."

He noted that "nitrites have been used as food preservatives for centuries."[53]

He also noted that aflatoxins may cause liver tumors.

"These compounds came to light," he said, "when thousands of turkeys died in England. Investigation showed they died of liver damage and that this liver damage resulted from toxins developed by a mold which grew on the ground nuts (peanuts to a large extent) which constituted a major portion of their diet. Wheat aflatoxin fed to trout produced liver tumors.

"Whether these aflatoxins are important to the human is as yet unestablished," he continued. "They have been detected in trace quantities in certain batches of peanut butter. Additionally, it is known that liver cancers have a peculiar distribution around the world, being rare in the United States but prevalent in the Chinese and certain African tribes."

One of the National Cancer Institutes research projects is the study of toxic products synthesized by molds that are intentionally added to the food for the purpose of fermentation. Moldy rice, prepared by growing certain strains of *Aspergillus* on rice kernels, and used in the Orient as a starter for fermentation of cereals and beans, has been shown to damage the livers of experimental animals.[54]

Researchers from the National Cancer Institute, the National Institute of Environmental Health Sciences, and the Akron, Ohio, Children's Hospital, have reported the activation of mouse leukemia virus by chemical carcinogens. One of the chemicals was urethane, used in agriculture, and another was diethylnitrosamine, a nitrite compound.[54a]

Even when cancer has been definitely linked to an additive, it takes a long time for that additive to be taken off the market. Red Number 4, used in maraschino cherries, was shown to be carcinogenic in animals, but is still on the market. Violet Number 1, used to mark federally inspected meat, and citrus red Number 2, used to make orange skins more attractive, have been found to cause tumors in animals. They have been sold under a "temporary extension" for more than eight years.[54b]

Proving a relationship between a chemical and cancer is extremely difficult today. Many people still do not believe the statistical evidence linking cigarette smoking and lung cancer. The problem is that cancer may take as long as twenty years to develop.

As Dr. Lijinsky said, "We are swimming in a sea of carcinogens."

Certainly, every additive to food should be studied for

its cancer-causing potential. Since most are added for economic purposes only, the consequences may not be worth it for one out of four of us.

ALLERGY

Allergy may not be as frightening as cancer or birth defects to most people, but to some who suffer from it, it can be just as deadly as any malady of man.

Take the case of a young man who had a stuffy nose and recurrent asthma attacks because of allergies. It was known that he was allergic to iodine and peanuts. One day, he took his wife out for ice cream. He ordered a banana split with chocolate and strawberry ice cream. He took a few delicious bites, gasped, grabbed his throat and turned blue. Without speedy medical attention, he might have died.

The young man, it turned out, was "violently sensitive to artificial Chocolatin and Strawberrin the two flavoring agents in the ice cream."[55]

This was a true case reported by Dr. Stephen Lockey, of Lancaster, Pennsylvania, an allergist who has been calling for recognition of allergic reaction to hidden drugs for years. He often quotes an Englishman, Dr. A Kewick, who wrote: "The world is short of food, and any substance which can be added to preserve food and yet be harmless to the consumer should be encouraged. Unfortunately, for all concerned, many additives considered harmless have proven otherwise."

Dr. Eugene Cowen, another allergist, wrote in the *Annals of Allergy*: "The attention to date on the problem of additives has been focused primarily upon the carcinogenic potentials, hematopoietic effects, intestinal bacterial flora changes and other physiological alterations. Except in a few cases, search of the literature has revealed no interest in the potential allergenic effect of excipient vehicles and additives."[56]

Unlike Dr. Lockey and Dr. Cowen, many scientists feel

that you cannot abstain from using a food additive just because some people are allergic to it.

However, Dr. Howard G. Rapport, writing in the *Journal of Asthma Research*, September, 1967, said: "Allergy is the most important chronic disease of childhood. Allergic illness is responsible for more days of school absence than any other chronic condition. It is the cause of loss of school, play time and growth time. It destroys healthy family life.

"The problem of allergic disease in childhood is unfortunately growing more significant each year. The question of what we are able to do about it does not have an easy answer. Every day we are exposed to new types of plastics, synthetic detergents, food additives, insecticides, weed killers and dozens of new chemical compounds which frequently end up in our rivers and reservoirs and the very air we breathe.

"Allergic illness kills more children each year than poliomyelitis, rheumatic fever or pneumonia.

"Each new compound is, of course, a potential invitation to allergy. New techniques in the production and storage of food have also taken their toll, permitting and even promoting early weaning and total dependence on manufactured formulas and introduction of solid foods as early as the first or second month of life. Allergically, this has created many problems."[57]

Dr. Cowen emphasized in his article: "Elimination of allergenic agents, where possible, has always been the cornerstone of the therapy for the allergic patient; with the increasing complexities of today's environment, the ability to adequately eliminate proved ingestant allergens is being seriously impaired. Much of this is the result of astronomical numbers of additives to our foods and drugs, and it is further complicated by the inadequate labeling of these additives, vehicles and excipients as to their exact nature and source."

Take the case of the "yellow peril." Sensitivity to aspirin has been known for more than half a century. It can cause fatal swelling of the tissues, stuffy nose, or asthma.[58]

Max Samter, M.D., of the University of Illinois College of Medicine, and Ray F. Beers, Jr., M.D., from the same institution, reported their findings in "Intolerance of Aspirin" in the May, 1968, issue of the *Annals of Internal Medicine*. They found certain patients showed symptoms of aspirin allergy after eating fresh pork, sweet corn, soft drinks, and cheese crackers.

"None of the patients who reported reactions after ingestion of food gave positive skin reactions to the food under suspicion," they wrote. "Analysis of the suspected foods established the presence of multiple additives in all but two of the samples. Of the preservatives, sodium benzoate was used most commonly; of coloring matter, hydrazine yellow 'tartrazine' [yellow # 5] turned out to be a component of all but one of the yellows in orange foods. The ability of hydrazine yellow to induce angioedema [tissue swelling] and respiratory symptoms in susceptible patients has been previously described but it had not been shown that reactions to hydrazine yellow and to aspirin co-exist in the same patients."

Since tartrazine and aspirin are chemically not alike but produce the same symptoms in aspirin-sensitive patients, the Illinois Medical College researchers conclude that the only thing they have in common is that they are minor analgesics. Whatever it is in the body that makes an aspirin-sensitive person react to the drug the way he does also makes him react to tartrazine, a hidden additive in food.

As a result, an aspirin-sensitive person who carefully avoids aspirin-containing products could have an asthma attack or serious edema after eating a food containing yellow coloring, and never know why.

PACKAGING

Allergy, cancer, and mutagenicity may occur not only from unintentional additives like pesticides and intentional additives like flavoring, colors, and surfacants, but also from packaging.

Paul E. Johnson, writing on the "Health Aspects of Food Additives" in the *American Journal of Public Health*, said: "There are several hundred chemicals used in formulating food packaging materials which might become additives; the basic glass, film paper, metal, cloth or wood plus the sizing and coating material; the plasticizers, the adhesives, the dyes and printing inks and the solvents, germicides, antioxidants and other miscellaneous chemicals associated with them."[59]

The Food Additives Amendment of 1958 gave the Food and Drug Administration the responsibility of making sure that indirect additives in foods, including those that migrate into them from packaging materials, be regulated to assure human safety. FDA Bureau of Science Director William H. Summerson told the National Conference on Indirect Food Additives in February, 1968, that his agency must be told for each new packaging material:

1. Whether or not it gets into the food.
2. How much may get in.
3. Whether that amount is safe for people.

Summerson noted that there has been great demand for using organo-tin compounds in packaging material. He said that the compounds have been shown to be extremely toxic and that only the FDA requirement of proof of safety prevented them from being used by the manufacturers. Unfortunately, the FDA and the food manufacturers cannot always predict what the consumer is going to do on his own.

Take the case of a fifty-five-year-old physician. He was hospitalized with an enlarged liver, progressive erythropenia (loss of red blood cells). He told his doctors that he had been feeling generally fatigued for quite a while with a sensation of heaviness in both arms, insomnia, headaches, loss of appetite, nausea, and occasional diarrhea. He had lost ten pounds in one month, and also had stomach cramps. This successful, wise, middle-aged physician presented a typical picture of lead poisoning. But how was he being poisoned?

In discussing his habits, his physicians learned that every evening the doctor sipped chilled soda from a mug made by his son. He usually drank two bottles of cola over several hours. He had been continuing this ritual for two years. The son had made the mug at college, glazing it with lead oxide. He had intended the piece for a vase, but his father preferred it as a mug. The soda leached enough lead from the mug's inner surface to poison the physician.

The doctors who made the connection between the mug and the lead poisoning also discovered that the ceramics industry uses some 25,000 to 30,000 tons of lead a year "and [that] there is a lack of government regulation for what could be a potentially dangerous situation."[60]

Another unexpected food additive may be from plastic dishes. Dorothy C. Smith, of the National Health and Welfare Service, Ontario, Canada, was asked whether formaldehyde might mix in with hot food when it is served on plastic dishes made from urea or melamine resins. She answered: "These tests show that hot liquids do, in fact, extract free formaldehyde from these plastics. Trace quantities were found in new, glossy dishes, but appreciable quantities were obtained when broken pieces were exposed to solvents at 57 degrees over a period of 24 hours. Under normal conditions of use, no significant amounts of formaldehyde would be released in food or drink from this type of plastic ware."

This is one woman's opinion, of course, and certainly the effects of formaldehyde released from plastic dishes mixed with certain foods and their additives should be studied further.

Many packages are made of plastic. Plastics behave differently with various foodstuffs. A plastic container that is not attacked by sugar or acid may be fine for jams. But the unsuspecting housewife may decide to reuse the empty container to collect drippings, unaware that certain toxic constituents may be extracted by the fat.[60a] In 1971, it was discovered that plastic bags containing blood for transfusions released a toxic substance into the blood. It was an unexpected finding, happened upon by chance by a Baltimore researcher tracking down a substance in the blood of patients. If someone were to look, perhaps he too would find similar problems with plastic food containers and wrappings.

THE LAW

The practice of adding chemicals to food is a very old one, says the Food and Drug Administration publication

"What the Consumer Should Know About Food Additives": "No doubt it began when man first learned to preserve meat by putting salt on it." The booklet goes on to say that during the early days of the Industrial Revolution in England and America, there was much trial-and-error experimentation with material used to preserve foods or to conceal inferiority by coloring them with dyes that were sometimes highly poisonous. Dr. Harvey W. Wiley's battle for the original pure-food law was to a large extent a fight against chemical preservatives such as boric acid, formaldehyde and salicylic acid.

"Food and Drug Administration scientists, appraising the situation in the mid-1950s, knew that several hundred of those additives were being used. They knew also that some of those in use had not been thoroughly tested for safety. Hearings were held by Congressman James Delaney which led to the amendments named after him. Under the law, prior to September 1958, the FDA could not prevent the use of a chemical simply because it was questionable or had not been adequately tested. It was necessary to be able to prove in court that the chemical was poisonous or deleterious. This is not difficult for chemicals which cause immediate or acute illness.

"But today, the big problem which concerns modern scientists is the long-term effect—what may happen in the body as the result of years or even a lifetime of exposure," the booklet said.[61]

Now the law reads that the food and chemical manufacturers are required to run extensive animal feeding tests on these additives before they are marketed. Results of these tests must be submitted to the FDA. If the FDA scientists are satisfied that the additive may be used safely, a regulation will be issued permitting its use.

The situation is certainly better than it was before the 1958 amendments. But are we really protected?

The American Medical Association's Council on Foods and Nutrition pointed out in 1961: "The new Food Additives Amendment gives special consideration to additives used in food prior to Jan. 1, 1958. Continued use of these additives is now permitted without toxicity tests if qualified experts generally recognize them as having been shown safe through either toxicologic tests or experience based on common use in food.

"Some chemicals used in foods had been evaluated prior

to the enactment of the Food Additives Amendment and found to be safe. These are allowed at the present time by virtue of this prior sanction. However, experience based on common use of food will not prove the absence of chronic harmful effects. Both animal test and experience with use are needed for evaluating safety. The Council on Foods and Nutrition believes that decisions to continue the use of additives should be based on demonstrations of their safety through scientific methods. Furthermore, the food industry should be allowed adequate time to demonstrate the safety of new chemicals used in foods."

The Council goes on to say: "The formulation of state laws for food additives should be encouraged to provide adequate regulations for all foods, whether or not they enter interstate commerce. Federal agencies do not have jurisdiction over foods that are produced and sold locally.

. . .

"At present, not only the laws but also the laboratories and inspection services in most states are inadequate to conduct a food control program comparable to that of federal agencies. Sufficient state funds should be appropriated to support research and testing on present food additives.

"Public and private funds should be appropriated to support research and toxicological testing of food additives. Only when sufficient scientific knowledge is obtained will we be : le to make progress in the expanded use of safe food ad itives."[62]

But recognition of the dangers of a food additive is very slow. When the Federal Food and Drug Act was passed in 1906, seven synthetic colors were approved for use in foods. By 1950, that number had grown to nineteen. In 1956, the government banned Food and Drug Certified Yellows numbers 1, 2, 3, and 4 and Red Number 1. In early 1965, it prohibited Red Number 4, but a few months later relaxed its ruling on Number 4 at the request of food and chemical companies. It is still in use.[63]

Coumarin, a component of vanilla-like flavoring preparations, was also banned after years of use. Coumarin has been shown to cause marked growth retardation, testicular atrophy, liver damage, and death in animals.[64] Safrole, a flavoring material, was also banned after years of use. It was found to cause testicular atrophy, cirrhosis, necrosis, and cancer of the liver in animals.[65]

The list is long—and growing longer all the time—concerning additives that have been used for years and then found to be harmful.

The Canadian Food and Drug Directorate in October, 1967, banned the use of noridhydroguaiaretic acid (NDGA), a food additive used to keep fat from going rancid. The move was made after animal studies showed that the antioxidant "may result in hazards to health."

NDGA is still being used in the United States in shortenings, cake mixes, and soft drinks, including Tang. The FDA has taken NDGA off the "generally recognized as safe list."[66]

Another problem, of course, is the illegal use of additives. Cakes seized in one of New York City's largest bakeries contained toxic dyes. Food and Drug inspectors seized seventy specialty cakes from the establishment, which did a huge catering business. The laboratory report showed that the cake and its decorative roses contained titanium dioxide, lead chromate, and dinitroaniline, ingredients of floral paint sprays.[67]

The FDA inspectors seized a shipment of cottage cheese in New Mexico containing coumarin, the liver-damaging, testicular-atrophying component of vanilla previously mentioned.[68]

With the limited number of Food and Drug inspectors who have to police millions and millions of shipments of food and drugs, chances of catching illegal additives are slim. Fortunately, most food and chemical manufacturers are honorable.

Former Food and Drug Commissioner James L. Goddard said: "In a society where we use three and a half pounds of food additives [per person] per year, we have to be re-evaluating constantly. We need more sophisticated testing programs; ones which give the consumer the assurance of safety he wants. We also have a growing awareness of the interaction of drugs."[69]

In the publication "Some Considerations in the Use of Human Subjects in Safety Evaluation of Pesticides and Food Chemicals," members of the National Academy of Sciences and the National Research Council showed they were well aware of the problem of food additive safety:

"A continuing problem facing industry and the public is that of establishing that a particular use of a chemical in food production, processing, packaging or storage will not be harmful to the consumer. It is generally recognized that absolute assurance that a usage will not prove harmful in any degree is unattainable. Therefore, effort is directed toward assuring that the hazard associated with the use is very small in relation to the health, agricultural and economic benefits to be derived."[70]

When they say "economic benefit," of course, it must be decided whether they are talking about the producer or the consumer. For instance, at the American Chemical Society meeting in April, 1968, a paper was presented on the properties of 1, 4, 5, 6-tetrahydro-2-acetopyridine. What is 1, 4, 5, 6-tetrahydro-2-acetopyridine? It is "A Cracker Odor Constituent of Bread Aroma." The authors of the paper said that, unfortunately, the flavor aroma of bread, which has universal appeal, fades within twenty-four hours after baking. They felt that the chemicals that smelled like bread could be used to spray a loaf and keep it smelling fresh.[71]

If the sixteen additives now added to bread makes it look and feel soft and fresh, and if 1, 4, 5, 6-tetrahydro-2-acetopyridine makes it smell fresh, is the bread fresh?

A&P is now advertising Jane Parker White Bread: "We date our Jane Parker White Bread. It's the only absolute guarantee of freshness you have ... unless you bake your own."[72]

How right they are! And we must ask if it is necessary to add chemicals to something to make us think it is fresh.

The National Academy of Sciences says: "As methods of investigation improve, it should be possible to improve the assurance of safety. Usually this assurance has rested on the interpretation of the results of exposure of experimental animals to the chemicals in question. While this approach has generally served admirably for assurance of safety of pesticides and food chemicals, its weakness is that the result obtained with laboratory animals may not always make it possible to predict what will occur in man.

"Once a chemical is released to the general environment, it would probably not be possible to identify exposed and non-exposed segments of the population or to classify persons by degrees of exposure, except to the extent that several groups might be identified whose dietary or other habits result in exposure different from that of the population.

"The possibility of carcinogenicity is particularly important, since the period of induction of cancer may be a considerable portion of the lifespan."

The Academy confesses that there is, as yet, no solution to the problem of translating toxicological data derived from animal studies into levels that provide an adequate margin of safety in human use. They also recognize that certain common disorders, particularly those affecting the kidneys, liver, and lungs, as well as chronic alcoholism, may greatly modify the response to and excretion pattern of many agents.

They suggest that if there is sufficient reason, once studies in normal volunteers have been completed and the safety limits for an agent established, similar studies might be carried out in selected groups of volunteer patients with disorders that may affect their responses. Whether manufacturers would take the trouble or whether patients with liver disease or chronic alcoholism would submit to the studies might be a problem. At least, an attempt should be made.

Besides patients with ailments, many scientists—including Dr. Walter A. Compton, President of Miles Laboratories, believes that the FDA fails to take into account personal likes and dislikes and eating habits. Dr. Compton made the statement in defending vitamin supplements, but it is just as true for other chemical additives.

Until we have better methods of testing, there are commonsense rules that would give us better protection from potentially harmful—even deadly—chemical additives. Alastair C. Frazer, speaking on "Unexpected Toxic Effects" in "Some Considerations in the Use of Human Subjects in Safety Evaluation of Pesticides and Food Chemicals," the National Academy of Sciences publication, suggests:

1. Epidemiological studies in those occupationally exposed. Any chemical substance proposed for introduction into the environment has to be manufactured, prepared, packed, or otherwise distributed. Those handling the substances are commonly exposed to it in quantities considerably in excess of those likely to reach the public at large. Possible toxic effects should be readily detected in such groups if appropriate studies are made.

2. Detailed investigation of accidental poisoning can also provide valuable toxicological information. This may be particularly useful in the case of substances with rela-

tively high toxicity for which special handling precautions are normally used. Accidents may occur either during manufacturing, distribution, or application. The use of trained combat teams to make an effective study of such incidents has been suggested.

3. With substances of low toxic potential, such as a new food additive, some control of unexpected effects might be afforded by restricting the initial marketing to a particular zone. Since this is not infrequently done for sales purposes, for example, to link television advertising with the distribution and display of new products, there might be some possibility of zoning distribution for safety purposes.

Lamenting the long, unwise use of DDT and cyclamates, geneticist James A. Crow wrote in "A Lethal Legacy," published by Field Enterprises in its *Science Year*, 1971: "It would be revealing to take a closer look at the incredible number of other man-made chemicals now found in our air, food, water, and even our medicine cabinets. Some of these, such as drugs and food additives, we consume on purpose. Others like sulfur dioxide and nitrates, we take in helplessly in our polluted air and water. But we know little of how any of these chemicals affect us, because persons living today are the first to have ever been exposed to many of them. These chemicals threaten us in four ways: they can poison us, cause cancer, deform our unborn children in the womb, and damage our hereditary material. . . .

"In the past, we have been quite reckless in our ignorance. New chemicals have been widely used long before much was known about their long-term effects on either man or wildlife. In my view, many should have learned an important lesson from the cyclamate episode. Never again should we add a substance to the diet of an entire nation without first performing exhaustive tests to determine its potential for both short-and long-term harm to human beings."

3 NOTES

1. Case of Rita L. Don, M.D., 102 University Towers, 1900 N. Oregon St., El Paso, Texas. Given to author by Steven S. Lockey, M.D., 60 N. West End Ave., Lancaster, Pa. 17603.

2. *New York Times,* March 4, 1968.

3. James L. Goddard, M.D., Food and Drug Commissioner, tape-recorded interview with author, May, 1968.

4. Howard J. Sanders, Associate Editor, "Food Additives," *Chemical and Engineering News,* October 10, 1966.

5. James L. Goddard, M.D., Food and Drug Commissioner, *Science News Letter,* December 10, 1966.

6. Orville Withers, M.D., *The New Physician,* July, 1968; correspondence with author, March, 1968.

7. *National Institutes of Health News Record,* May 28, 1968.

8. Stephen Lockey, M.D., 60 N. West End Ave., Lancaster, Pa., correspondence with author, March, 1968.

9. Howard J. Sanders, *op. cit.*

10. *Ibid.*

10a. William B. Mead, United Press International, February 22, 1971.

11. T. C. Byerly, "Use of Human Subjects in Safety Evaluation of Food Chemicals," National Academy of Sciences, 1967.

12. *Federal Register,* April 11, 1968.

13. H. R. Bird, Ph.D., Department of Poultry Husbandry, University of Wisconsin, *American Journal of Clinical Nutrition,* May–June, 1961, Vol. 9.

14. *Federal Register,* April 11, 1968.

15. *Physicians Desk Reference 1968,* Medical Economics, 22nd edition, p. 1015.

16. Bernard Robins, M.D., "Deterioration of Renal Function Due to Tetracycline," *Beth Israel Hospital Journal,* 1964.

17. Murray C. Zimmerman, *Archives of Dermatology,* January, 1959, Vol. 79.

18. T. C. Byerly, *op. cit.*

19. Frank Rosen, allergist and former president of the New Jersey Allergy Foundation. Correspondence with author.

19a. Konrad Wicher, Ph.D., Robert E. Reisman, M.D., and Carl E. Arbesman, M.D., "Allergic Reaction to Penicillin Present in Milk," *Journal of the American Medical Association,* April 7, 1969, Vol. 208, No. 1, p. 143.

20. *Medical World News,* May 17, 1968.
21. W. C. Hueper and W. D. Conway, *Chemical Carcinogensis and Cancers* (Springfield, Ill., C. C. Thomas, 1964), p. 654.
22. H. R. Bird, *op. cit.*
23. *Ibid.*
24. Associated Press reports from Washington, D.C., May 6, and May 25, 1971.
25. J. C. Roe, *Current Concepts in Cancer, JAMA,* May 13, 1968, Vol. 204, No. 7, p. 142.
26. Hueper and Conway, *op. cit.*
27. News from the National Institutes of Health, December 15, 1967.
28. American Medical Association release, August 28, 1968.
28a. Samuel S. Epstein and William Lijinsky, *Nature,* January 3, 1970.
28b. "Food Preservatives, Medicines Seen Possible Cancer Agents," Associated Press, December 14, 1970, Washington, D.C.
28c. John T. Litchfield, Jr., M.D., "Drug Toxicity in the Human Fetus and Newborn," *Applied Therapeutics,* September 1967, pp. 922–926.
29. Barry Commoner, speech before the American Association for the Advancement of Sciences, New York, December 27, 1967, and speech before the National Parks Association, Washington, D.C., March 16, 1968.
29a. John W. Olney and Lawrence G. Sharpe, *Science,* October 17, 1969, p. 386.
29b. Edward A. Arees and Jean Mayer, "Monosodium Glutamate—Induced Brain Lesions; Electron Microscopic Examination," *Science,* October 30, 1970, pp. 549–550.
30. Hueper and Conway, *op. cit.*
31. Hearings before a subcommittee of the Committee on Government Operations, House of Representatives, June 10, 1970, Washington, D.C.
32. *Ibid.*
33. W. David Gardner, in *The New Republic,* September 14, 1968.
33a. Subcommittee of the Committee on Government Operations, *op. cit.*
34. *Ibid.*
34a. *Ibid.*
34b. *Ibid.*
34c. *Ibid.*
34d. Senator Warren Magnuson, in his letter to Secretary of Health, Education and Welfare Elliot L. Richardson, December 22, 1970.
34e. *Medical Tribune,* July 6, 1970, p. 11.
34f. *Ibid.*
34g. Samuel S. Epstein, Children's Cancer Research Foundation, Boston, Massachusetts; Alexander Hollaender, Oak

Ridge, Tennessee; Joshua Lederberg, Stanford University; Marvin Legator, Howard Richardson, Food and Drug Administration, Washington, D.C., and Arthur H. Wolff, Consumer Protection Environmental Health Service: joint letter to *Science,* December 26, 1969, p. 1575.
35. *Agricultural Research,* March 16, 1968.
35a. *Medical World News,* January 22, 1971, p. 13.
36. *Modern Medicine,* January 1, 1968.
37. R. E. Eckhardt, M.D., Ph.D., Esso (New Jersey) Research Medical Bulletin, 26:142, July, 1966.
38. *New York Times,* May 19, 1968.
39. "Do Chemicals Sow the Seeds of Genetic Change?" *Medical World News,* April 26, 1968.
40. *Ibid.*
41. *Medical World News,* May 17, 1968.
42. *Ibid.*
43. *Ibid.*
43a. "Additive Is Found of Genetic Import," *New York Times,* June 28, 1971.
44. *Ibid.*
45. Dr. Philippe Shubik, "The Cancer Bulletin" of the Chicago Medical School, August, 1967. Dr. Shubik has moved to the University of Nebraska.
46. William Lijinsky, M.D., Division of Oncology, Chicago Medical School, "The Cancer Bulletin," August, 1967.
47. *Health Insurance News,* March, 1968.
48. Morton Levin, M.D., Assistant Commissioner for Medical Services, New York State Congressional Hearings on Food Additives, 1957.
49. Law 85–929, December 6, 1958.
50. *Journal of the American Medical Association,* November 18, 1961.
51. Color Additives Amendment, p. 17.
52. "Problems in the Evaluation of Carcinogenic Hazards from the Use of Food Additives," Food Protection Committee, Food and Nutrition Board, National Academy of Sciences, December, 1959.
53. R. E. Eckhardt, M.D., *op. cit.*
54. *Progress Against Cancer* (National Institutes of Health, 1966).
54a. Howard J. Igel, Robert J. Huebner, Horace C. Turner, Paul Kotin, and Hans L. Falk, "Mouse Leukemia Virus Activation by Chemical Carcinogens," *Science,* December, 1969, Vol. 166, pp. 1624–1626.
54b. United Press International reports, January 4, 1971.
55. Case of Stephen D. Lockey, Lancaster, Pa.
56. Eugene Cowen, M.D., Assistant Clinical Professor of Medicine, University of Colorado Medical Center, *Annals of Allergy,* March, 1967, Vol. 25.
57. Howard Rapport, M.D., *Journal of Asthma Research,* September, 1967, Vol. 5.

58. Max Samter, M.D., and Ray Beers, Jr., M.D., "Intolerance to Aspirin: Clinical Studies and Consideration of Its Pathogenesis," Annals of Internal Medicine, May, 1968, Vol. 68, No. 5.

59. Paul E. Johnson, "Health Aspects of Food Additives," American Journal of Public Health, 56:6, June, 1966.

60. Robert W. Harris, M.D., and William R. Elsea, M.D., "Ceramics Glaze as a Source of Lead Poisoning," Journal of the American Medical Association, November 6, 1967, Vol. 202, No. 6.

60a. Linda Grace, "There Are No Harmless Substances," World Health, April, 1969, pp. 20–22.

61 "What Consumers Should Know About Food Additives," FDA publication 10, revised in June, 1962.

62. "Safe Use of Chemicals in Foods: A Council Statement," Journal of the American Medical Association, November 18, 1961, Vol. 178, No. 7, p. 749.

63. Howard J. Sanders, "Food Additives," Chemical and Engineering News, October 10 and 17, 1966.

64. W. H. Hagan, et al., Division of Toxicological Evaluation, FDA, Food and Cosmetics Toxicology, April, 1967, Vol. 5, No. 2.

65. Ibid.

66. Paul Schvette, Deputy Assistant Commissioner for Education and Information, FDA, HEW, personal correspondence June 5, 1968.

67. Abraham Goodman, Assistant to the Council President of New York City, personal communication with author, November 21, 1968.

68. FDA Papers, February, 1967, p. 35.

69. Tape-recorded interview, May, 1968.

70. "Some Considerations in the Use of Human Subjects in the Safety Evaluation of Pesticides and Food Chemicals," Food Committee, National Academy of Sciences, 1965.

71. Irving R. Hunter, et al., "Preparation and Properties of 1, 4, 5, 6-tetrahydro-2-acetopyridine, a Cracker Odor Constituent of Bread Aroma," American Chemical Society meeting, San Francisco, April, 1968.

72. A&P advertisement for Jane Parker Bread, 1968.

4

Meat and Poultry Hazards

As all good shoppers know, when buying meat you should consider its color—a bright red means freshness. However, sodium nicotinate and related substances act to preserve the bright red color of meat. In many states it is legal to use these additives. One company that produces the additive purportedly produces enough to doctor 320 million pounds of meat a year solely for cosmetic and therefore for economic purposes.

The Council on Drugs of the American Medical Association in 1960 recommended that the intake of nicotinic acid be 4 milligrams daily for infants, 13 to 17 milligrams daily for those between thirteen and twenty years of age, and 12 to 18 milligrams for adults.

When a doctor's wife became ill after eating hamburger meat, the irate physician investigated and found that their butcher had added a handful of sodium nicotinate to their hamburger despite the fact that instructions on the chemical container said to add only two ounces to one hundred pounds of meat.

One evening, members of a sorority house at Northwestern University sat down to a dinner that consisted of Swedish meatballs, noodles, broccoli, butter, rolls, green olives, celery, chocolate roll, coffee and sugar and cream. There were glasses of iced water at each place setting. The boys hired to wait on the tables carried in the heavy trays, laden with food for 121 persons. It was an ordinary meal on an ordinary school night until ten minutes after it began. Then one of the girls stood up and yelled, "I'm burning up! I feel like I'm in an oven." A blond-haired girl with a light complexion, she looked as if someone had splashed red paint over her face. Her skin burned and itched, and her ears tingled. She complained of a pain in her stomach, and began to shake with chills.

Within seconds, thirty-nine of the girls and five of the male student waiters developed the symptoms. They had extreme flushing, tingling, itching, dry mouth, stomach pains, and swelling of the face and knees. Oddly, the worst symptoms were among the fair-haired members of the group.

Doctors summoned to the sorority house found blood pressures ranged from 80 to 160 systolic and 58 to 80 diastolic, and pulses varied from 64 to 96. Something the students had eaten had caused dilatation of their blood vessels and a severe allergic-like reaction.

The doctors began administering Benadryl to the students, as antihistamine, at 25 to 50 milligrams a shot. Either the Benadryl ran out or some students refused to take it, but within twenty to forty minutes the symptoms began to subside in students whether they had received the medication or not. That is, the symptoms eased in all but two students. They had to be hospitalized.

What was it that had caused the sudden illness? Why did only forty-four out of the 121 persons eating the same food become ill? Why were some more severely affected than others? Doctors had to check food storage, preparation, food service, pesticides, rodenticides, utensils (were they galvanized or cadmium lined?) and disinfectants. Nothing was found to account for the poisoning.

The following morning, despite initial denial, it was discovered that at the meat-purveying firm sodium nicotinate had been added to the ground beef before it was delivered to the sorority house. Ironically, chemical tests that were performed at the Chicago Board of Health on some of the ground meat proved negative. Later, however, microbiological tests showed as much as 225 milligrams of sodium nicotinate per 100 grams of meat.

The officials of the meat firm confessed that two ounces of powdered sodium nicotinate had been sprinkled on the ground meat directly following the first coarse grind and prior to the second fine grind. Because of the uneven distribution of the chemical and because of the varying susceptibility, only forty-four out of the 121 became ill.

Similar stories could be told about incidents concerning

sodium nicotinate around the country. Eight-eight persons in 38 families became ill over a thirteen-day period when the butcher from whom they purchased meat began using the chemical.

The doctors who described these cases in "Food Poisoning Due to Sodium Nicotinate" in the *American Journal of Public Health,* October, 1962, were Dr. Edward Press and Dr. Leona Yeager. They said: "In the opinion of the authors, the use of such additives should be properly controlled and regulated in order to avoid misuse. The products in which it has been used should be properly labeled or some other method of making the consumer aware an additive has been added should be utilized."[1]

The above cases concern a relatively mild illness with a legally, if improperly, used additive.

But the truth of the matter is that even with the new meat legislation in 1968, you can still literally buy a pig's eye in your frankfurter. In other words, there may be a lot more in your meat than you bargained for.

Rodney E. Leonard, Deputy Assistant Secretary of the Consumer and Marketing Service, United States Department of Agriculture, said in testimony before Congress in 1967:

"There exists considerable disparity between statutory provisions of federal, state and local laws which creates a form of economic separation that carries with it significant advantage for the unregulated. . . .

"To be more specific, excessive water can be used in the processing of hams and other pork cuts; cereal, flour and dry milk can be substituted in the manufacture of sausage. Such chemicals as niacin and sodium sulfite can be used to prevent color fading and mask spoilage. Products such as frankfurters and bologna can be labeled 'all meat' when in fact they contain all types of meat by-products, extenders, excessive water and fat.

Oscar Sussman, D.V.M., M.P.H., president of the New Jersey Public Health Association, charged in the February 19, 1971, issue of the American Public Health Association's *Nation's Health*:

"The consumers of the United States have grown up. They should no longer be treated as babies, or imbeciles. True facts should be given to them in all cases with regard to foods, drugs, and the environment. When given all the facts we can depend on our consumers to properly evaluate them and take action to protect themselves.[2]

"The American Public Health Association has adopted a policy covering meat and poultry inspection. This policy in essence states that the APHA believes that U.S. Government labels and concurrent advertising that imply to the public that meat and poultry, when 'U.S. Inspected' or 'U.S. Inspected for wholesomeness,' are 'guaranteed safe,' leads the consumer into a false sense of security. It is a known fact that, of samples taken of U.S.-inspected meat and poultry products, as high as 50 percent or more contain Salmonella and other pathogenic organisms that may have caused food poisoning when improperly handled. That when consumers now see the U.S. Guarantee of implied safety, they do not properly handle the product, relying on their government's false assurance of total safety. This applies equally to tapeworm in beef, trichinosis in pork, and other pathogens on raw meat and poultry."

Dr. Sussman said the APHA intends to go to court unless the government makes public the possible hazards still contained in "U.S. Inspected" meat. He said APHA wants labeling on meat packages similar to the warnings of cigarette packages.

In 1967, after much debate, Congress passed the Wholesome Meat Act. It required the Agriculture Department to take over meat-inspecting duties in states that did not develop, by December 15, 1970, a program "equal to" federal meat-inspection standards. Only intrastate inspection programs were affected because meat sold interstate was already under federal inspection.

As of January 11, 1971, 24 states were certified equal. One, North Dakota, was taken over by the U.S. Department of Agriculture's inspection program, and 28 were still under USDA consideration—meaning that they had not yet met the standards four years after the law passed. (Texas and West Virginia were certified in April, 1971.)[3]

A Congressionally ordered investigation of meat processing plants by the Government Accounting Office in 1970 turned up unsanitary conditions in 41 out of 48 plants. Of the 48, 40 were USDA-inspected plants.[4]

Senator Abraham Ribicoff (D–Conn.) immediately called for full-scale investigation of the Agriculture Department. He cited the report as a "shocking illustration of the need for more vigorous enforcement of the Wholesome Meat Act."

Examples of the conditions found and reported to the Congress on June 24, 1970, by General Accounting Office investigators were:

FEDERALLY INSPECTED PLANT B

During the period June, 1968, through October, 1969, Consumer and Marketing Service, Department of Agriculture, inspectors made eight reviews of this plant. On October 29, 1969, a Government Accounting Office investigator accompanied Consumer and Marketing Service inspectors and found, among other things:

"There were several areas, especially the slaughter, offal, and inedible departments, where the walls, columns, and door and window casings were broken, peeling, and crumbling. Several window panes were missing, and windows were not screened. . . .

"In the offal department, meat scraps from the previous day's operations were not removed from floor and equipment. A few fresh hams and uncured bacon slabs in the offal cooler were contaminated with granular or flaked material. Condensation from a refrigeration component was dripping on some of the products. Bone chips were found in pork brains. . . .

"In the pork cooler, a carcass had fallen to the floor and was being splattered with water by an employee washing the floor. Carcasses contacted the floor occasionally while being transported to the pork cut department.

"In the inedible product area, crud, dead cockroaches, and a decomposed rat were observed. Also a strong odor hung generally throughout the area and permeated one's clothing while passing through. Some of the containers were not clearly marked to show that they were to be used only for inedibles.

"Product packaging material was contaminated by mouse droppings and a putrid piece of meat. Concrete pillars being contacted by beef carcasses were not metal clad to facilitate cleanliness. Sewage frequently backed into plant areas. Unwrapped frozen product was stored on unclean wooden racks.

"Some doorways and openings provided accessibility to rodents. Evidence of rodent runs was observed along the foundations of the plant and outer buildings and a few dead cockroaches and a live mouse were observed inside the plant.

"During the review we observed metal shavings protruding from the blade of a carcass-splitting saw. We easily removed some of the shavings from the blade which, in our opinion, could have become embedded in a carcass and could have resulted in a hazardous product. We showed the metal shavings to the C&MS reviewer but he did not require that the blade be replaced. . . ."

NONFEDERALLY INSPECTED PLANT G

"Rat feces were observed throughout the plant, including beef-boning areas and carcass coolers.

"Rats had chewed holes in the wooden cooler doors.

"Rats' nests and a birdnest were observed in the plant.

"A live rat was observed in the tank house.

"Moldy meat scraps were accumulated behind a refrigerator unit in the cooler. Two packages of moldy meat were lying on a boning table.

"Work tables and equipment were dirty. In general, floors, walls, ceilings, and loading docks were dirty.

"Open doors and windows were not screened."

Senator Ribicoff stated: "We must find out why the department [of Agriculture] is unable to carry out its directive in the supervision and regulation of federal meat quality standards."

There were 7,500 veterinarians and food inspectors—including intermittent and part-time workers—checking the 3,200 USDA-inspected meat plants in 1971.[5]

The Nixon Administration turned down a task-force recommendation that a new consumer agency be set up to handle meat and poultry inspection. The report cited some of the problems in administering current inspection programs:[6]

● Federal meat and poultry inspectors—the people who actually check items on a plant's production line—are supervised by veterinarians, who in turn report to higher-ups, all the way up to the U.S. Department of Agriculture executives. As a result, reports sometimes get lost, and there are strained relations between food inspectors and veterinarians in many areas. In some plants, the veterinarians have lost control of the inspectors.

● There is a "very inbred staff" in the consumer protection hierarchy, and inspectors and veterinarians vie for promotion and exclude other people without similar backgrounds.

In general, the GAO report referred to the lack of authority and responsibility at many levels of the system, which detract from efficiency of consumer protection programs.

As in the case of the FDA, the USDA is not protecting us against unwanted chemicals in our food. The Department of Agriculture, in 1970, slashed by 74 percent the number of meat samples checked for antibiotic residues and by 50 percent the number of tests for residues of the hormone, diethylstilbestrol. As a result, only one out of every 124,000 cattle carcasses were checked for DES. Evidently, to make up for the lack of testing, the government began the first prosecutions of farmers who ship livestock contaminated with DES in 1971.[7]

The USDA was called upon to defend itself when it was discovered by the press that nearly 103,000 cattle carcasses checked by federal inspectors during the 1969/70 fiscal year had "cancer eye" or similar tumorous disorders. The carcasses were held in the meat plants until the tumors were whittled out and then sent to market.[8]

Dr. Joseph S. Stein, head of slaughter inspection for the Department, said that cutting away an infected part from a red meat animal is no worse than a housewife removing a spot from an apple and using the rest of it for a pie.

The USDA tried to get away with the same thing in chicken. In 1970, the Department announced it would allow chickens bearing cancer virus disease to go on the market as long as the chickens did not look too repulsive. The affected organ or part would be cut out, just as in beef. After a great public outcry, the Department had a change of heart and said that it would continue its policy of keeping cancerous chickens from reaching American dinner tables.[8a]

A poultry inspection law similar to the meat law was

passed in 1968. State poultry inspection programs had until August 18, 1971, to develop standards of inspection equal to the federal standards. As of January 11, 1971, 5 states have been certified equal; 13 states have been designated federal—meaning they have no hope of developing a suitable program of their own—and 36 states have been granted extensions to try to develop programs.[8b]

Poultry, by its very nature, is contaminated with bacteria on the surface. The more it is handled, the more germs grow and the more likely it is to make you sick.[9] Clean birds or parts packed in ice or held at 32 to 34 degrees F. should have a shelf life of approximately four days without a marked rise in surface contamination. If the temperature is allowed to rise even a few degrees, serious contamination will occur. If birds are held at room temperature for only a few hours before being placed under chilled temperature, chances of causing illness are increased. If the surface of the skin or exposed flesh is slick, watch out! It is likely that the bacteria count on that piece of poultry is in the millions.

In January, 1968, federal inspectors discovered that one out of every five nonfederally inspected chickens was not fit to eat. They found that the "errors" in nonfederally inspected poultry included passing products with "gross lesions of disease as well as failure to remove infectious processes and contamination of the body cavity with stomach contents or fecal material."

The Agriculture Department disclosed in 1971 that samplings of poultry for illegal residues of organic arsenic turned up positive in one-fourth to one-sixth of the birds tested. Organic arsenic is far less toxic than inorganic arsenic. It is added to poultry feed to make the birds grow faster. Federal regulations permit one part per million of arsenic in the livers, giblets, and other organs of poultry after slaughter and 0.5 parts per million in muscle tissue.[10]

Senator Joseph Montoya, of New Mexico, in his statement before the Subcommittee of the Committee on Agriculture and Forestry of the United States Senate, in 1967, eloquently explained why more federal legislation of meat and meat products was necessary:

"Products ready for home use without further preparation have proliferated. They are more highly processed, with greater opportunity for adulteration and deception that a purchaser would not notice.

"All these advancements complicate inspection procedures. Inspection of meat for wholesomeness, without use of chemical analysis, is most effective when meat is fresh. Frozen meat must be defrosted prior to inspection. If there is suspicion of adulteration or unwholesomeness, it must be examined and tested closely, often requiring laboratory analysis.

"Adequate inspection of a fast processing line is far more difficult than is visual examination of carcasses and fresh meat. Also in modern processing, there is often less opportunity for inspectors to check each stage of operation."

Senator Walter F. Mondale (D–Minnesota), who worked to get through the 1967 Wholesome Meat Law, pointed out: "Modern science has developed and continues to develop new wonder chemicals—additives, preservatives and colorations—which can prevent the consumer from using the normal smell and sight tests to determine spoilage or deterioration, which will also provide a crude consumer test for toxic or diseased meat or meat products growing alongside the spoilage bacteria. Consumers have no chance at all to determine the safety of the meat they buy."

The first Meat Inspection Act, March 18, 1907, was put into force after the author Upton Sinclair revealed conditions in meat plants in his book *The Jungle*.

The new meat bill passed on November 21, 1967, provides matching funds for states to improve their meat-inspection programs. It also:

● Extends the federal program to commerce wholly within the District of Columbia or within any territory not having a legislative body.

● Prohibits commerce in animal products not intended for human use unless denatured or in animal products properly identified as not intended for human use or naturally inedible.

● Provides for record keeping by certain slaughterers and handlers; registration and regulation of certain handlers of dead, dying, disabled, or diseased animals.

● Authorizes regulation of meat storage and handling to prevent adulteration or misbranding.

● Provides for withdrawal or refusal of inspection service, for detention, seizure and condemnation, injunction and investigation as new enforcement tools.

● Subjects meat imports to the same requirements as those for domestic meats.

● Otherwise revises and clarifies the Federal Meat Inspection Act.

In a letter to the author, Senator Mondale said: "I am totally satisfied with the new [meat] law, and I feel that it was the best possible bill that could have been passed."[11]

But, unfortunately, there is still plenty of opportunity—although certainly not so much as before—to slip adulterated meat onto the consumer's plate.

The fact remains that federal inspection is no guarantee of clean meat today. The majority of meat has been inspected federally for more than sixty years, but hog cholera and brucellosis are so prevalent in America that other countries, including England and Sweden, have banned the import of pork products from the United States. There is more trichinosis in the United States than in any part of the world. About one in six inhabitants of this country harbor the trichinae, with clinical symptoms displayed by 4.5 percent.[12]

United States Department of Agriculture inspectors grade fresh cuts of meat in descending order of quality: Prime, Choice, Good, Standard, Commercial, or Utility. Retail stores most often carry meat of "Choice" quality. The three upper grades indicate meat of excellent quality, flavor, and tenderness. The three lower grades are better for braising, pot roasts, stews, and so on.

Once a plant is in operation, every live animal is supposedly examined in its holding pen by a veterinarian or by an inspector supervised by a veterinarian. After slaughter, federal meat inspectors are supposed to examine each carcass, including internal organs and glands, to be sure that no unwholesome condition exists. A carcass that passes this inspection is stamped with a purple USDA shield mark.

During the production of processed meat, inspectors are supposed to examine both meat and nonmeat ingredients, and check a dozen or more steps to ensure that meat is wholesome and prepared according to approved formulas. They inspect curing solutions, check processing times and temperatures, and control the use of restricted ingredients.

They must see that all containers are properly filled and labeled.

Sounds good on paper, doesn't it?

In New York City, a city health inspector walked into a *federally* inspected *kosher* sausage and frankfurter manufacturing firm on the lower east side in February, 1968. "Federally inspected" means that a USDA inspector had to certify the meat. "Kosher" means "clean" and prepared under the supervision of a rabbi trained in such work. The New York City inspector found approximately seventy-five violations, and noted, among them:

"The worm gears in the meat grinder were rusty and caked with bits of old fat and meat. Paint was scaling off the equipment and falling into the hot-dog mixtures. Fresh meat was being stored in rusty tubs.

"A sterilizer required in Federal plants to contain 180-degree water for setrilization of knives that are dropped on the floor was full of cold, greasy water. A dead roach floated in the scum of the water's surface.

"Evidence of rats was everywhere, even where meat was being handled . . . and there was a Federal inspector on the premises."[13]

Then there was the "Case of the Counterfeit USDA Meat Stamp." One must marvel at the officers of a wholesale meat company in New York for their brazenness. During 1963, the Feldman Veal Corporation of 410 West Thirteenth Street and two of its officers were charged with selling inferior, ungraded meat bearing counterfeit Department of Agriculture seals to the United States Military Academy at West Point. The company shipped 24,330 pounds of veal legs to the Academy for about $16,000. It was supposedly "Choice" cut.

One of the company's competing firms became suspicious of the low bids. A losing company has the right to inspect the product of the winner. This was done; and, according to Jack Kaplan, Assistant United States Attorney, it was discovered that the shipped meat was one and two grades inferior to Choice meat and that the USDA seals were counterfeit.[14]

The Feldman company received a $1 fine in Federal Court after conviction. Abraham E. Abrahamson, head of the New York City Department of Food and Drugs, declared: "It is always our hope that when we win a case in the public interest, there will be a suitable penalty."[15]

But even the best law is no protection against unscrupulous operators and dealers in dirty and diseased meat.

Dr. M. R. Clarkson, then Associate Administrator of the Agriculture Research Service, United States Department of Agriculture, said before the Food and Drug and Cosmetic Law Section of the New York Bar Association, in New York City: "One of the toughest jobs in meat inspection—the task of certifying the wholesomeness of processed meats—grows increasingly difficult as new techniques are used . . . and as housewives depend more and more on the processor to prepare foods once cooked at home.

"We can't depend on laboratory analysis of samples to protect consumers against unfit or adulterated meat and poultry products. It is impossible to detect them—even microscopically—when mixtures have been chopped or blended or cooked."[16]

His comment becomes even more appalling when you read the National Livestock and Meat Board's booklet "Facts About Sausage." It states: "With over 200 varieties of sausage and ready-to-serve meats to choose from, menus built around them are literally endless.

"When selecting sausage or ready-to-serve meats, choosing some of the less familiar ones can add excitement to the menu. . . . "[17] Like Russian roulette?

There was a hue and cry by consumers when the USDA permitted frankfurters to contain chicken and still be called "all beef." But chicken is the least of what can be found in some processed meats.

In 1968, Consumers Union found insect fragments and larvae, rodent hairs, and other impurities in 6 of 48 samples of federally inspected, fresh pork sausage. In 1970, they found insect fragments and rodent hair in some of the sausage used by the Campbell Soup Company for their Swanson frozen breakfasts.[17a]

Consumers Union pointed out that the USDA enforces the Federal Meat Inspection Act, and the FDA enforces the Food, Drug, and Cosmetic Act. The USDA is responsible for the meat that goes into sausage, but rarely inspects the spices because it considers that the FDA's job. The FDA tolerates a certain level of filth in many foods, Consumers Union charges. The organization points out that the American Spice Trade Association tolerates various imported spices that contain between 0.5 and 2 percent, by weight, of extraneous matter, including stones, dirt, wire, string, insect parts, and animal hair. Spices in this country, according to CU, are never inspected by the FDA unless they have been processed or stored in unclean facilities.

Lack of enforcement, according to Consumers Union, allowed contaminated sausage to get on the market.

As for the fat content of hot dogs, the President's Consumer Affairs adviser, Virginia Knauer, pressured through a 30 percent fat-content ceiling for frankfurters. Frankfurter makers contended that Americans wanted a juicier hot dog. Back in 1958, the famous Brussels World Fair frankfurter—especially concocted to show off the tastiness of the American dish—contained 19.8 percent fat.[17b]

There is room for economic fraud with processed meat. When New York City's Consumer Affairs Commissioner Bess Myerson ordered an inspection of the city's eating places, it was found that 23 percent were serving meat dishes that did not meet city or state standards.[17c]

Of the 180 establishments visited by inspectors during two days in 1970, 42 were accused of having served hamburger made with cereal, or of selling veal patties advertised as veal cutlets.

However, it is the harmful bacteria and chemicals that are of more concern.

Neil Peck, former Assistant United States Attorney, Southern District of New York, has pointed out the difficulty in identifying bad meat, even under the federal inspection system. In an article written in cooperation with James C. G. Conniff for *Pageant* magazine, in May, 1968, Peck said that he began an investigation into violations of the Federal Meat Inspection Laws in New York City in 1964. It required more than fourteen months to complete:

"This unhealthy meat, which came from various parts of the country, had been derived from so-called 4-D sources. It had been frozen into 50-pound blocks and then shipped east to the New York metropolitan area. In its frozen state—dark as a mahogany table top, so that you couldn't tell by looking at it what was in it . . . the shipment of frozen meat blocks was put directly into the machines that manufacture these processed meats."

Peck said inspectors can spot-check only one or two cartons in an order containing 30,000 to 40,000 pounds of meat. If such a shipment goes to a plant with two federal inspectors, it would take them weeks to inspect the meat properly in that one shipment. They would have to melt down all the blocks to make sure it was all good meat.[18]

Improper use of legitimate chemicals is another hard-to-detect area of meat inspection.

Dr. K. E. Taylor of the United States Department of Agriculture Meat Inspection Service, writing in the *Journal of the American Veterinary Medical Association,* in 1962, said there are approximately twenty insecticides, eleven feed additives, and seven therapeutic agents that have been approved by the Meat Inspection Division. "However, if they are not used properly, even they leave a residue in the meat derived from cattle, sheep, calves, swine and goat."[19]

But how can you tell? Even the Meat Inspection Division's own regulations and instructions pertaining to biological residues in meats say: "In acute poisoning resulting from cholinesterase inhibiting pesticides, the lesions are never pathognomonic or particularly outstanding."[20]

As far as antibiotic residues are concerned, if the injection is not given in the usual place, the haunch, inspectors may miss the fact that the animal has been loaded with antibiotics just before slaughter. At best, even federal inspectors can just spot-check for signs that the animals have been injected recently.[21]

Although laboratory tests may not reveal some forms of contamination, they certainly are necessary. The visual inspection of meat alone, without chemical and biological testing, is as old-fashioned as crossing the ocean in a sailing ship.

David C. Tudor, V.M.D., Research Professor of Poultry Pathology at Rutgers, said in an article in *Veterinary Medicine,* in February, 1967:

"A reliable poultry disease diagnosis is based upon clinical and laboratory findings from several living and dead birds that typify the flock condition. It is essential that all parts of the bird be examined for gross abnormalities or evidence of disease."

Here are some of the laboratory procedures Dr. Tudor believes necessary for adequate poultry inspection:

● To establish the kind and extent of intestinal parasitism, fecal scrapings from four regions of the intestinal tract are examined under low-power microscopy.

● To establish rapidly the presence of bacterial infections such as tuberculosis and erysipelas, impression smears of the spleen and liver tissues are sometimes helpful. An acid stain must be prepared to identify tuberculosis organisms.

● Evidence of anemia, leukemia, or parasitism may be gained from blood smears prepared from the heart, blood, or blood taken from the brachial wing vein and stained with Wright's stain.

● Fowl cholera may be diagnosed when paradimethylamino benzaldehyde solutions added to the broth culture of fowl cholera yields a pink color. In the absence of this diagnostic tint, a diagnosis of the disease cannot be made.

● To identify salmonella infection, initial bacterial agar colonies may be transferred to nutrient peptone or tryptose broth and then to S.S. agar or MacConkey's agar.

In addition to the laboratory tests, there are about two hundred clinical observations that Dr. Tudor says can lead to a diagnosis of disease in fowl. Such diagnosis, of course, is difficult to make when birds go past an inspector—even a federal inspector—at the rate of approximately six a minute on an assembly line.[22]

As for the laboratory inspection of meat, the necessity was made clear in *Recommended Methods for the Microbiological Examination of Foods,* edited by J. M. Sharf and published by the American Public Health Association in 1966:

"Meat appears on the market in many forms and microbiological methods are needed to assay for quality, detect spoilage and determine the source of any quality defects. As a result of various methods of producing, processing and handling meat products, some meat foods act as selective or enrichment media for specific types of microorganisms."[23]

Some states have little or no facility for testing food for adulteration and contamination. A large percentage of food inspectors have no degrees beyond high school, and some are close to illiterate.

Under the present system, even a federal meat inspector may be assigned to the same plant for years. During this time, he is bound to become very friendly with the personnel. In other investigative branches of government, inspectors are periodically moved about to prevent such temptation.

But perhaps the biggest flaw in the protection of meat, even with the new act, is the lack of properly trained scientists and inspectors. Money cannot buy skilled employees who do not exist.

WHAT CAN YOU DO TO PROTECT YOURSELF FROM ADULTERATED FISH, MEAT, AND POULTRY?

In answer to the above question, most health officers to whom I spoke said, "Nothing." They explained that there is no way, in today's marketplace, for a consumer to be sure that the fish, meat, or chicken she is buying is wholesome. Sight and smell are no longer to be trusted because the chemicals that are in use today can doctor decaying, diseased, or just plain stale food. All a consumer can do, they say, is to rely on the federal, state, or local inspector, and put his faith in the brand name of the product and the store selling it.

However, it is apparent to me, after engaging in research for this book, that there are things that consumers can and should do to protect their food. Here they are:

1. Report any sanitary violation found in a store and any contamination of packaged food discovered at home.

2. Report any outbreak of suspected foodborne illness in the family. (While most boards of health and other protective agencies may be too understaffed to search out abuses effectively, they usually are only too happy to follow through on a suspected offense.)

3. Urge any group you belong to or do-it-yourself, to find out how your state is implementing fish, poultry, and meat inspection. Some states are sure to claim insufficient funds or insufficient personnel, and to lag behind on food-inspection improvements.

4. Don't purchase any meat in a supermarket if its container has been damaged in any way.

5. Don't buy an unrefrigerated canned ham. Canned hams must be refrigerated at all times, and many supermarkets ignore this cardinal rule.

6. Don't contaminate the meat at home yourself. Fresh

meat should be unwrapped as soon as it comes from the market. Store fresh meat uncovered or loosely covered in the coldest part of the refrigerator. Cured meat should also be stored in the refrigerator. Canned hams should be kept under refrigeration. Frozen meat should be stored at a temperature of O degrees F. or lower. It may be placed in the refrigerator under ordinary refrigeration if it is to be used immediately after defrosting. Never refreeze meat.

7. Do not leave frozen fowl unrefrigerated to defrost. Keep them in the refrigerator even if it takes several days for them to thaw.

8. Don't place other food in a container that previously held raw meat, and do not place other foods on a counter on which raw meat, fish, or poultry has been without first disinfecting the counter with soap and water.

9. Wash your hands after touching raw meat.

10. Practice good hygiene. Wash your hands after going to the bathroom, and don't handle meat when you are ill with a contagious ailment, especially an infected finger.

11. Never, never eat raw or undercooked meat.

12. Don't look for bargains. If meat, fish, or poultry in any form is being sold far below the normal price, there's a reason, and it is usually not for the benefit of the consumer.

4 NOTES

1. Edward Press, M.D., and Leona Yeager, M.D., "Food Poisoning Due to Sodium Nicotinate," *American Journal of Public Health,* October, 1962.
2. Oscar Sussman, D.V.M., M.P.H., "Lullaby of U.S. Inspections," *The Nation's Health,* February, 1971.
3. William Bloom, Information Division, United States Department of Agriculture, Washington, D.C., personal communication with author, January 12, 1971.
4. "Weak Enforcement of Federal Sanitation Standards at Meat Plants by the Consumer and Marketing Service," Report to the Congress by the Comptroller General of the United States, June 24, 1970.
5. William Bloom, *op cit.*

6. "New Meat Inspection Unit Idea Vetoed by Nixon Aide," Associated Press, Washington, D.C., October 16, 1970.

7. "U.S. Moving to Sue Farmers for Trace of Drugs in Cattle, United Press International, Washington, D.C., February 8, 1971.

8. "U.S. to Review Meat Policies," Associated Press, Washington, D.C., April 10, 1970.

8a. "Meeting Planned on Poultry Check," Associated Press, Washington, D.C., January 26, 1970.

8b. William Bloom, *op. cit.*

9. *Recommended Methods for the Microbiological Examination of Foods*, edited by J. M. Sharf (American Public Health Association, 1966), p. 119.

10. "Arsenic Is Found in Poultry but It's 'Not Dangerous,' " Associated Press report, Washington, D.C., January 7, 1971.

11. Personal correspondence, February 14, 1968.

12. World Health Organization Report, 1957.

13. *New York Times*, March 4, 1968.

14. New York Criminal Court, Judge Milton Shalleck, June 18, 1968.

15. "City Health Official Is Critical of $1 Fine in Sale of Dirty Veal," *New York Times*, July 3, 1968.

16. M. R. Clarkson, Associate Administrator, Agriculture Research Service, USDA, before the Food and Drug and Cosmetic Law Section of the New York Bar Association, New York City, January 24, 1962.

17. "Facts About Sausage," National Livestock Meat Board, 36 S. Wabash Ave., Chicago, Ill.

17a. "Sausage and Spice," *Consumer Reports*, June, 1970, p. 344.

17b. "Chickening Out on the Frankfurter," *Consumer Reports*, January, 1970, pp. 32–33.

17c. "Eating Place Study Finds 10 Percent Serving Shamburger Here," *New York Times*, February 1, 1970.

18. Neil Peck, former Assistant U.S. Attorney, Southern District, New York, as told to James C. G. Conniff, *Pageant*, May, 1968, Vol. 23, No. 11.

19. K. E. Taylor, D.V.M., USDA Meat Inspection Service, *American Veterinary Medical Journal*, November 15, 1962.

20. "Biological Residues in Meat and Meat Food Products," Instructions and Standards of Compliance, 11:27, November 7, 1964.

21. Ruth Winter, Newark *Star-Ledger*, March 5, 1965.

22. David Tudor, D.V.M., Research Professor of Poultry Pathology, Rutgers, in *Veterinary Medicine*, February, 1967.

23. *Recommended Methods for the Microbiological Examination of Foods*, edited by J. M. Sharf (American Public Health Association, 1966), p. 111.

5

A Fish Story

Because a three-year-old boy of Haddon Heights, New Jersey, was hungry a little earlier than usual, his mother cooked some fillet of flounder for him; the rest of the family was to eat their portion of the fish later. The little boy gobbled it down, ate his dessert, drank some milk, and then announced that he was going to his grandmother's, nearby.

The rest of the family sat down to eat their dinner after the boy had left.

The child had just reached his grandmother's when he complained of feeling ill. He turned blue and began gasping for air. An ambulance was summoned, but before it could speed its tiny burden to the hospital, the little boy died.

His parents and the rest of the family, all of whom were in the midst of their meal when they heard about the boy, stopped eating. But they began to suffer from dizziness, nausea, and cyanosis (lack of oxygen).

Later that same evening, the Poison Control Center in Philadelphia reported to the Chief of the Philadelphia Milk and Food Sanitation Section that three women who had eaten at a Philadelphia restaurant were hospitalized when they became ill within an hour after their meal. All three women had eaten flounder. All three had become faint, suffered headaches, dizziness, and vomiting, and were cyanotic.

Robert C. Stanfill, District Director of the Food and Drug Administration, and a neighbor of the boy's family

117

was shaving when he heard a report of the youngster's death on the radio. He left immediately for his office.

He and physicians and other inspectors who became involved in the case concluded that a chemical which looks like ordinary table salt but which is illegally used to freshen rotting fish, sodium nitrite, was probably the cause of the poisoning. A chemist was assigned to run tests for nitrites on portions of the foods eaten by the victims.

Stanfill recalled: "Meantime, Philadelphia Department of Public Health Inspectors, New Jersey Department of Health personnel, the Camden County New Jersey Detective Bureau personnel, Haddon Regional Health Commission and many other local officials were investigating reports of illnesses and were sampling fish, searching for records of sales of sodium nitrite and otherwise participating in individual and group efforts to obtain all the facts.

"The facts pouring in definitely showed the presence of large quantities of sodium nitrite in the fillets delivered to the Yeadon chain-store warehouse and the one restaurant sale, but fillets sold to other stores by the same and other fish dealers did not show such evidence," he continued. "These findings narrowed the probable distribution of toxic fillets. Evidence of nitrites was present under the brine hoops on the concrete floor, and on the cutting table of Universal Seafood Company, Inc. The president and employees denied any use of nitrites. The proof was not yet definite as to how the incident occurred."

The health departments and the federal inspectors involved in the case issued public warnings concerning the possibility that people might have in their possession contaminated fillets of flounder.

"About the same time that Daniel DiOrio [president of Universal Seafood] was on a Philadelphia television program denying—as a spokesman for the fish industry—that anyone in the fish industry had or would use sodium nitrite, Dr. John Hanlon, the Philadelphia Director of Public Services, and I appeared on Station WRCV-TV after 11 P.M. to assure the public that the distribution of the poisoned fish didn't go beyond the nearby 3-state area around Philadelphia," Stanfill said.

"It was on Good Friday that the persistence of Food and Drug Inspector Frederick A. Cassidy resulted in the manager of a chemical and salt supply house joining Cassidy in a search of the firm's sales records covering a period of more than a year. The records showed a hurry-up early-morning delivery of 400 pounds of sodium nitrite had been rushed to Universal Seafood Co., Inc., on the Monday before the little boy died on Tuesday night. A smaller shipment had been delivered during Lent the year before. Investigation disclosed that DiOrio had personally authorized the order and personally accompanied the 400-pound drum from the truck to the filleting room. The drum bore bold red letters."

Stanfill told his colleagues at a Food and Drug meeting that the defendants, DiOrio and Universal Seafood Company, pleaded "no defense" to a charge alleging addition to fillets of a "poisonous and deleterious preservative for the purpose of misleading and defrauding."

He said the persistent investigational work by Food and Drug inspectors, police officers, health officers, public health inspectors and agents received recognition. But he pointed out that the quiet, efficient work done in the laboratory also deserves commendation. Stanfill described how the laboratory rapidly developed the proof that the fillets had been adulterated with nitrite to give them the smell and appearance of freshness when the fish were far from it.

"The facts were presented to a Federal Grand Jury," Stanfill recounted. "The Grand Jury indicted the fish processor, Universal Seafood Company, Inc., of Philadelphia, Daniel DiOrio, its president and treasurer; and Noel LoCastro, its foreman. . . . The corporation which went out of business soon after the poisoning was publicized, was fined $100. DiOrio was fined $2,500 and sentenced to a year in jail, with 11 months suspended during a 3-year probation period. DiOrio's attorney petitioned the court to release DiOrio from jail and presented medical testimony that DiOrio had a heart disease and imprisonment was not good for his health. DiOrio was released after 16 days."

Stanfill said: "I am usually objective about matters pertaining to sentencing and serving sentences. Many citizens were indignant that DiOrio had only a month to serve and were more so when he was released after 16 days. . . .

"When DiOrio was in court for sentence after pleading 'Nolo Contendere,' a parade of fish dealers and other business associates and competitors appeared as character witnesses. A spectator remarked: 'But for the grace of God, it may have been any one of them.'

"I could not fully agree, but it seemed to be a sad commentary that a man who deliberately poisoned rotten fish, and whose deliberate action caused a child to die, was wholeheartedly supported by his fellow merchants. After dissolving Universal Seafood Co., Inc., DiOrio has continued his lucrative fish business in his own name."[1]

The amount of nitrite used as a preservative in the fish that was eaten by the little boy was enough to kill him quickly, but there is a good possibility that smaller amounts of nitrite added to fish may kill still others, more slowly. Present law permits the addition of nitrates and nitrites in concentrations of 500 and 200 parts, respectively, to one million parts of fish or meat. On March 3, 1971, in an article in the *Journal of Agricultural and Food Chemistry*, a publication of the American Chemical Society, five FDA chemists reported "potent cancer agents called N-nitrosamines are present in certain fish that have been treated with ordinary food preservatives consisting of nitrogen compounds called nitrates and nitrites." The chemists were Thomas Fazio, Joseph N. Damico, John W. Howard, Richard H. White, and James O. Watts. They called the N-nitrosamines one of "the most formidable and versatile groups of carcinogens [cancer agents] yet discovered" and said that there is a growing apprehension among experts about the use of nitrites and nitrates. N-nitrosamines are formed when nitrates and nitrites combine with amine, a chemical naturally present in the stomach.

One of the problems is that, twenty-five years ago, there were relatively few types of fish products offered for sale compared with what is available today. In those days, the problems of ensuring freshness and preventing decomposition of fish products were paramount, according to J. Kenneth Kirk, Associate Commissioner for Compliance of the Food and Drug Administration.

Kirk said: "Recent years have seen a marked increase in the production of 'convenience' foods where the consumer has to do no more than heat the product to serving temperature. In the seafood area, this includes breaded shrimp, tuna potpies, breaded fish sticks and portions, crabmeat, deviled crab, etc. The elaborate and frequently untried production methods involved in processing of such foods have introduced new problems. These items require a great deal of handling in the producing plant and consequently plant sanitation is of prime importance to insure a safe and wholesome finished product."[2]

Such new developments, however, are not the only problem. They compound old problems—lack of inspection and outdated American methods of fishing and fish-product production. These problems are growing worse, not better, because of competition from foreign imports.

There were, in 1971, 62 fish inspectors from the U.S. Bureau of Fisheries assigned to 40 plants, a number that is hardly adequate to cover the 4,946 fish processing plants and 75,000 fishing vessels.[3]

From 1959 to 1969, there were 16 known deaths and more than 2,500 cases of serious illness from the consumption of fish products; in 1970, 25 patients died in a Maryland nursing home reportedly after eating a meal of shrimp and eggs.[4]

Consumers Union has tested fishery products through the years. In 1961, CU found 75 percent of the fried fish sticks substandard. In the same year, it found 63 percent of the breaded shrimp tested substandard or not grade certified. In 1963, 46 percent of the frozen fish fillets were found substandard. In 1965, 50 percent of the raw fish portions tested were substandard, and in 1966, 98 percent of the salmon steaks tested were substandard.[4a]

What does substandard mean? Fish sticks that earn the Grade A seal from the U.S. Department of the Interior must have good flavor and an odor typical of the fish and of the breading, with no rancidity, staleness, bitterness, off-flavor or off-odor. They must also score at least 85 out of a possible 100 points in 13 other quality factors.

In Consumers Union's view, bacteriological standards are a must for frozen foods. When it tested frozen fish sticks in 1970, if found that 51 percent of the samples had enterococci (fecal streptococci).[4b]

"How do fish sticks become contaminated by feces?" Consumers Union asked rhetorically. "There's no way to tell. Sewage dumped in the ocean could contaminate the fish even before they are caught; contaminated water could be used to make the ice that keeps the catch cool on the trawler; the water used to rinse the fish at the processing plant could be contaminated by a toilet or sewer line;

the workers who handle the fish during processing could neglect their personal sanitation.

"But no matter how the contamination occurs, it is the responsibility of the fish processor, the last link in the manufacturing chain, to make sure contaminated fish sticks stay out of the supermarket."

Self-regulation of the fishing industry has evidently not worked.

In the Senate hearings relating to the Wholesome Fish and Fishery Products Act of 1969, the following testimony taken directly from reports by Food and Drug inspectors was presented:[4c]

INSPECTION BY FDA SEATTLE DISTRICT, JANUARY 14-16, 1969

"Inspection of this smoked salmon operation disclosed sanitary conditions to be deplorable. Extensive rodent activity, of recent origin, was noted throughout the plant. Processing equipment was encrusted with heavy, dark, dried, discolored fish residues. Smoking racks and boxes were of wooden construction, records were improperly maintained, and management refused to code its product. Other objectionable conditions included poor employee practices, improper design of equipment, and improper processing techniques."

FRESH FISH PACKER AND FILLETER, MARCH 4, 1969

"This plant operates under primitive conditions. The community has no municipal sewage system and when the nearby river is high, water carrying raw sewage backs up in ditches near the plant. Doors and windows are not adequately screened.

"Approximately 2 dozen flies were observed around the fresh fish on the floor and table. This was unusual in that the day was cool enough to require heat. When the cesspool cover was lifted, I saw 8 to 10 roaches run for cover.

"The plant surroundings offer an excellent opportunity for contamination of the finished product. The environment in which the fish are dressed has both live and dead animals. A dog has free run of the plant. Live turtles are cultivated in an area adjoining the fish market. Dead turtles are not removed from the surroundings, but are permitted to remain and draw flies. Armadillo pieces are scattered about the turtle pit.

"A putrid odor comes from the turtle pit.

"The insects have access to the contents of the cesspool. The cesspool overflows, leaving standing water much of the time. The surrounding area has very poor drainage.
. . .

"Wooden boxes are used and reused to transport and store fresh fish. This is before and after dressing. These wooden boxes are haphazardly stored on the outside of the plant.

"Dirty cardboard poultry crates are used for collecting fish heads. These fish heads will be used for crawfish bait. The bait in the cardboard crates is stored along the side of the edible fish in the walk in the cooler.

"The food handlers have a dirty appearance, None wore head coverings. The soap dispenser was empty. VK Power and Pinezol are used as disinfectants.

"Fish scales and other material from the plant floor are swept through a hole in the wall to the outside. Here the material is permitted to stand for long periods of time."

SMOKED FISH PLANT, INSPECTION, JANUARY 14-16, 1969

"An estimated 100 large rat pellets were observed in the northeast corner of the processing area where the kippering racks and wood smoking poles are stored when not in use. This heavy rodent activity was within several inches of these food contact surfaces and would easily be exposed to rodent activity. This is especially significant when considering the management admitted that this equipment (cooking racks and poles) is never washed or sanitized.

"Hundreds of rodent pellets both large and small were observed on and about a pile of old cardboard cartons in the warehouse. These cartons were covered with hundreds of pellets, heavy rodent urine stains, and nesting material.

"Heavy residues were noted on most of the processing equipment which was being used. These residues were primarily dried scum type material which appeared to have accumulated for quite some time. The wood framed cooking racks shown in photos #10 and 12 through 15 were encrusted with a heavy layer of fish material and fish type oil. Management admitted that their racks are never washed or sanitized. In fact, Mr.—— stated that the wood racks were so encrusted that he could obtain the desired smoke flavor without using any smoke in the cooler and the fish would pick up enough flavor from the wood racks.

"The knives used in the plant had wood handles or were cord wrapped which gave off a putrid odor. Many had rusty blades and were stored in a slot type rack against a dirty wall. . . . "

"In 1963, inadequate safeguards in processing smoked lake fish resulted in a major outbreak of botulism; nine persons died. During one weekend in 1966, about 400 cases of salmonella poisoning were reported. The illness was directly attributable to smoked fish.

"In 1966, more than 250 cases of food poisoning were

reported in one city; imported shrimp was identified as the cause.

"Infectious hepatitis has been linked to shellfish in four different incidents since 1964. A total of 309 confirmed cases of hepatitis have been traced to shellfish from four eastern coastal areas. These are some of the reported cases. No one knows how many cases of illness attributable to fish go unreported every year. Clearly, unless consumer protection methods are strengthened, further cases of disease or death may be anticipated."[5]

By piecing reports together, a picture can be obtained of the factors that caused the contamination and how they led to the illness of the unsuspecting fish eaters.

On May 30, 1966, a bar mitzvah was held in Edison, New Jersey. A bar mitzvah is an occasion marking the time when a Jewish boy of thirteen years has attained the age of religious duty and responsibility. There were 121 guests in all, arriving in two shifts—one beginning at 2:00 P.M. and the second at about 8:00 P.M.

According to Dr. Eugene J. Gangarosa and his colleagues who documented the case in the January, 1968, *American Journal of Public Health*,[6] the next day, May 31, fever and upset stomach were reported by many who attended the party. Their symptoms developed from one to seventy-two hours after the affair. They had diarrhea, abdominal cramps, nausea, vomiting, fever as high as 104° F., shaking, chills, dehydration, headache, and general prostration. The uncomfortable illness lasted from two to ten days.

The bar mitzvah party was catered by a Newark, New Jersey, firm, and smoked whitefish was among the thirty items served. The same menu, except for the smoked whitefish, had been served by this catering firm at four other gatherings, with no reported illnesses following. By questioning those who became ill, it was discovered that the one thing they had eaten in common was the smoked whitefish.

Samples taken from the ill guests showed the presence of *Salmonella* identified as *S. java*.

The smoked whitefish was purchased from a delicatessen at 11:00 A.M. on May 30 and was taken directly to the bar mitzvah party arriving at noon. *The delivery truck was unrefrigerated.*

"The delicatessen had received its usual large shipment

of smoked whitefish from its customary source, a fish-processing company in Brooklyn, New York, on Friday morning, May 27, 1966. The fish was stored in a large freezer which was in good operating condition. The delicatessen was open for business over the Memorial Day weekend, during which most of the whitefish was sold. None of the employees of the delicatessen had been ill prior to the Memorial Day weekend, but two of them who ate smoked whitefish developed the fever and upset stomach during this period.

Dr. Gangarosa said: "More than 12 separate outbreaks similar to the one bar mitzvah 'blitz' occurred during the Memorial Day weekend; more than 300 cases of fever and upset stomach were reported by local health departments throughout the greater metropolitan area of New York, New Jersey and Pennsylvania. Several patients required hospitalization, but there were no known deaths.

"The vehicle of infection was traced to one of the major whitefish smoking plants in the area. On April 1, the smoking plant had been seriously damaged by an extensive fire. Major plumbing and structural repairs were being made from the time of the fire until just before the outbreak. *During this time, there was some decrease in refrigerator and freezer space.* Careful investigation of the plumbing after the outbreak revealed no cross connections and no obvious source of sewage contamination.

"There was no known illness among the employees preceding the outbreak, although two individuals involved in the processing of salmon had fevers and upset stomachs simultaneously with the identical illness experienced in the outbreaks described above. Their stool cultures and the cultures of nine other employees were found positive for *S. java.*

"Although many denied it, it was generally accepted that almost all employees from the janitor to the manager ate processed products while on the job. Whitefish was especially favored; management accepted this as an inevitable part of the operation.

"Contaminated fish were voluntarily withdrawn from the market. The plant voluntarily ceased operation, and an extensive environmental survey was conducted a few days after the outbreak. Every phase of the operation was reviewed and numerous cultures were obtained. No evidence could be found of environmental contamination.

"Fish involved in the epidemic were caught in Canada. They were dressed, frozen and then shipped by truck to Winnipeg, where they were repackaged, glazed and stored until February, 1966. Thirty-one cartons of fish were shipped by rail to another company in Brooklyn which delivered four cartons to the smoking plant. On May 24, fish in three of these four cartons were responsible for the epidemic. *S. java* was isolated from one fish in the remaining carton obtained from the freezer of the smoking plant.

"Subsequently, two other shipments from the Great Slave Lake, one of which did not go through Winnipeg, were found positive for *S. java*. These fish had been caught and processed in June, 1965.

"Shipments of contaminated fish were traced to two companies at Hay River on Great Slave Lake. Thus the source of contamination was clearly at Great Slave Lake.

"The most likely cause of the contamination," Dr. Gangarosa concluded, "was probably the practice of using raw river water to wash fish after dressing and to make ice to pack the fish in. A combination of circumstances that favored the growth and proliferation of salmonellae at the processing plant probably caused the epidemic."

What happened when federal Food and Drug inspectors examined fifteen plants in one section of the United States engaged in the processing of smoked fish?

Fish and fish products are highly perishable, even more perishable than meat and poultry, which have received more attention from government agencies.

In 1967 and 1968, hearings were held by the Consumer Sub-committee of the Committee on Commerce of the United States Senate under the chairmanship of Senator Warren G. Magnuson. The purpose was to draw up a bill "to protect the nation's consumers and to assist the commercial fishing industry through the inspection of establishments processing fish and fishery products in commerce."

Testifying before that committee in July, 1967, Harold E. Crowther, Director of the Bureau of Commercial Fisheries, said:

"Fish and fishery products are highly perishable and must be given proper care during handling and processing. Over the years, the United States fishing industry has improved the quality of its products. State inspection, the Department's Voluntary Inspection Program, the Public

Health Service's Shellfish Inspection and actions of the Food and Drug Administration have contributed to this improvement.

"Much more needs to be done, however, if fish and fishery products are to enjoy a position of trust and confidence comparable to competitive meat and poultry products. It is a well-known fact that per capita consumption of fishery products has remained fairly constant over the past twenty years, while that of poultry and meat has increased substantially. It is our belief that uncertainty of quality is a major consideration in limiting consumer acceptance of fishery products. . . .

"We believe that the problem exists because of the lower-grade products being intermingled with these good products. This makes it difficult for the consumer to really determine what she is getting."

He was asked by Senator Philip Hart, of Michigan, who was also on the Senate Committee: "What special problems exist in fishery products as a food that are not present in other foods?"

Crowther said: "I believe one of the points I mentioned in my statement is that fish is even more perishable than meat and poultry and requires special handling from the time it is caught until the time it reaches the consumer.

"I think one of the problems we have with fishery products is that many of them are caught at sea. The fish die as they are brought aboard. And from that moment on they must be handled carefully and preserved in ice or by freezing and brought ashore.

"With meat and poultry products, an animal comes right to the plant itself. But from the time the fish does die, from that point on, it is highly susceptible to deterioration in quality. Therefore, I think it is a product that needs special attention."

He went on to say that because "we do catch fish in the oceans and the gulf and even in our inland waters, by necessity, many of our plants are scattered along the entire coastline of the United States. And in many respects they must be fairly small.

"Being small also presents problems, because you do not have a large organization behind you. Adequate financing is needed to make sure that the products can be held in proper storage until the market can absorb them.

"Because the plants are small and in many cases inadequately financed, I think many of them find it difficult to take all the precautions that must be taken to protect the fishery products adequately."

Senator Hart then said: "And the fellow who has the finest plant in the world is handicapped by the existence of the problem which your last answer implicitly suggests exists—namely, that once the fish is on the shelf the consumer doesn't know whether it is from the finest of all plants or the worst of all."

Crowther answered: "That is entirely true. For example, if we had 100,000 pounds of top-quality fish in one lot and 10,000 pounds of lower quality in another lot which were offered for sale at the same time, the poor quality fish would depress the price of both lots of fish. Also, the consumers who bought the poor quality product would be discouraged from buying any fish for some time.

"Unfortunately, in most segments of our industry, there is no premium placed on top-quality fish as landed by the vessel. This is not the fault of the processors, because in many cases, if they attempted to restrict buying from a particular fisherman because his quality was not up to par, in times when fish were needed this boat would not go to him. It would seek another processor."

The proposed Wholesome Fish and Fishery Products acts of 1968–1970 were aimed at:

● Developing a comprehensive federal program for consumer protection against the health hazards and mislabeling of fish, shellfish, and seafood products.

● Setting standards and developing continuous inspection and enforcement.

● Helping the states develop their own inspection programs.

● Assuring that imported fish and fishery products are wholesome.

The new fishery bill, like the others, would:

1. Require a nationwide survey of the operations and sanitary conditions of commercial fish-processing establishments and commercial fishing vessels for the purpose of developing effective standards of sanitation and quality control under which such establishments and vessels must be operated and maintained. These standards would be published by regulation. The survey must be completed and regulations issued within one year after funds are

appropriated for this purpose. The regulations would be effective no later than one year after publication, but if the Secretary deems it necessary the effective date can be extended for an additional year.

2. Once the regulations are effective, the owner of every fish-processing plant and commercial fishing vessel which does not provide fish or fishery products for use as human food solely for distribution within one State must certify that they are complying with the regulations and must obtain a "certificate of registration." If he fails to do so, he cannot operate.

3. Establishments of vessels that obtain the certificate and later fail to comply will have their certificate suspended until they take the necessary steps to comply.

4. The bill would require continuous inspection of each such establishment. This means that a full-time inspector will be assigned to each plant except in cases where geographic distribution and plant size make it possible for one man to inspect more than one establishment on a substantially continuous basis in a manner that would carry out the objective of this proposal. The processing of fishery products is, in many cases, seasonal, and plants are located in close-enough proximity to permit an adequate inspection on such a basis. Because of the large number of fishing vessels, it would not be practicable, nor is it necessary, to provide a continuous inspection of them. Continuous inspection does not mean that every fish, as in the case of meat, would be inspected. This is not practical, but the fish would be spot inspected while in the plant.

5. The bill would provide that fishery products processed for interstate or foreign commerce in an establishment with a certificate and packed in containers or wrappers shall be labeled.

6. The bill would make it unlawful to transport or sell fish or fishery products in commerce that are both capable of use as human food and are adulterated or misbranded when sold or transported.

7. The bill would authorize the issuance of regulations covering the conditions under which fish and fishery products capable of use as human food shall be stored or handled. These would apply to anyone buying, selling, freezing, storing or transporting or importing fish and fishery products in or for interstate or foreign commerce,

to prevent adulteration or misbranding when delivered to the consumer.

8. Once the regulations of the Secretary are effective, fish or fishery products cannot be imported into the United States if they are adulterated or misbranded or otherwise fail to comply with the provisions of the act or the regulations applicable to domestic fish and fishery products in commerce. Individuals who obtain these [products] for their own use are not subject to this requirement with a 50-pound weight limit.

"The bill also provides that if the Secretary determines that the country of origin has a system of plant, vessel and product inspection at least equal to the United States system, and that reliance can be placed on certificates required by United States regulations, and that the article complies with the system of the country of origin, such certificates may be accepted by the Secretary as proof of compliance with the comparable requirements of this act."

There are 116 nations involved in fishery trade with the United States. This includes our close neighbors, Canada and Mexico, as well as more distant places such as Barbados, Surinam, India, Pakistan, Senegal, Malaysia, and the Canary Islands. Over 50 percent of the edible fish in the United States comes from foreign countries, some of which have very good inspection programs and some none at all. In arguing against continuous inspection, during the 1969 Senate hearings, J. Ray Duggan, president of the National Fisheries Institute, said: "There are some who say the inspection bill providing for continuous inspection will serve as a trade barrier. This should not be allowed. If so constructed it would harm the U.S. consumer immeasurably.

"Imports of shell fish come from more than 70 nations, many of whom would be hard pressed to meet requirements of a continuous inspection bill due to lack of trained personnel and financial problems, but who with proper effort meet a more reasonable approach."[7]

Is it unreasonable to make sure imported fish is safe to eat?

In order to reduce the costs the proposal would enable the Secretary to limit, with the approval of the Secretary of the Treasury, the number of ports where the imports could come in.

Like the Meat Act, the bill would also provide a cooperative program under which the Federal Government could assist the States in developing and administering a State fish inspection program where the State has adopted a State mandatory inspection law which is at least equal to the Federal law. This would apply only to persons, firms, or corporations engaged in the State in processing fish and fishery products for use as human food solely for intrastate distribution.

James Ackert, president of the Atlantic Fishermen's Union, made the point that if foreign fishermen are not required to meet the rigid standards of United States fishermen, and if United States fishermen have to sustain the additional costs of inspections, then they will be in an even more disadvantageous position than they are today. Furthermore, the American consumer cannot tell whether he is buying foreign or domestic fish when he purchases certain fish products.

Mr. Ackert said the Atlantic Fishermen's Union had a laboratory test made on foreign fresh fillets on May 19, 1967: "There was evidence of an additive being used to extend the shelf life of the fillet, namely chlortetracycline [an antibiotic]." He pointed out that the fisherman is uniquely dependent upon a rapid sale of his fish and unloading of his vessel. In contrast to the operations of the beef industry, he cannot place his fish in stockyards for leisurely inspection.[8]

American fishing boats are antiquated. Unlike many European ships, they still use wooden pens for their catch of fish instead of a hard-surfaced material, such as aluminum, which is easier to clean.

According to Leroy Houser, of the Standards Development Section, Shellfish Sanitation Branch, Division of Environmental Engineering, Department of Health, Education, and Welfare:

"It is said to be important that the fish be stored in the ice [aboard ship] so that any water from the melting ice drains away and does not form a pool of blood and slime along the dorsal part of the cavity. Otherwise, the gut cavity soon develops sour off odors indicating spoilage to those trained in public health, so some means of flushing the gut cavity with chlorinated sea water or potable water should be necessary as a public health measure.

"Most fish in the North American fisheries are landed in

an uneviscerated form. However, extreme care is necessary in storing the fish in holds as the weight of the ice and fish may squeeze out intestinal material of fish stored in the lower levels. This material has been shown to contain up to 10 million bacteria per gram of gut content."[9]

When fish is stored from eight to sixteen days aboard an American fishing vessel, there are many opportunities for spoilage. A solution has been found by other countries in the form of floating fish factories. They compete with American fishermen.

The Soviet Union is building a huge, multipurpose factory ship. The 43,000-ton *Vostok* will serve as flagship to fourteen fishing vessels, operate a cannery capable of filling 150,000 cans a day, store 13,000 tons of bulk cargo, and provide quarters for 600 crewmen. These supertrawlers, the society says, are equipped with electronic gear to find cod, machinery to process catches, and giant freezers to preserve the fish. When the ships reach port, the fish are ready for the grocers.[10]

It seems that the only innovation American fishermen have been able to make is to cut down on the number of seamen aboard ship to try to reduce the cost of bringing in fish.

Another who spoke out against passage of the Fishery Products bill was Dr. George A. Michael, Director of the Division of Food and Drugs, Commonwealth of Massachusetts, before the Committee on Commerce, on April 24, 1968:

"Gentlemen, it is my considered opinion after twenty-seven years in the Division of Food and Drugs, eighteen years of which were served as director of that division, that the concept of continuous inspection in regards to any area of the food business is archaic, obsolete and it does not take into consideration the great advances made by the scientists in our federal and state laboratories.

"We can determine many more of the items that are a hazard to the consuming public in our laboratories than an inspector looking at a product can determine by visual examination.

"It is a waste of money to insist on mandatory, continuous inspection in any food field except in the slaughtering operations."[11]

Though the bill was shelved in committee, Senator

Warren G. Magnuson, chairman of the Committee on Commerce Hearings, said of the proposed Wholesome Fish and Fishery Products Act of 1968:

"Due to the difficult problems raised by several provisions of this bill and the late date at which it was introduced, it now appears that the Commerce Committee will not report out a fish inspection bill this year. We anticipate, however, that a similar bill will be introduced early in the next session of Congress, and I suspect that some legislation will be enacted at that time."[12]

With increasingly polluted water and wider distribution of fish, it is imperative that the American consumer be protected against disease-bearing or low-quality fish.

Many communities discharge human excrement into convenient rivers, lakes, or estuaries. It was always assumed that the contaminating material was naturally purified by dilution, aeration, exposure to sunlight, and microbial action. However, a researcher from the Biological Sciences Laboratory at Fort Detrick, Frederick, Maryland, Werner A. Janssen, and a researcher from Chesapeake Biological Laboratory of the Natural Resources Institute, University of Maryland, Caldwell D. Meyers, found this was not true. They found that live fish may become infected with dangerous germs from man, and thus constitute a public health problem. They tested white perch from the rivers that flow into Chesapeake Bay through heavily populated areas.[13]

Researchers from the Department of Microbiology, University of New Hampshire, tested samples of seawater and oysters collected during periods from April to November, 1964, 1965, and 1966. They isolated a number of different *Salmonella* serotypes from such samples of seawater and oysters in shellfish-growing areas that met the current recommended standards for approved shellfish-growing waters. They commented: "It is apparent from these results that the current bacteriological criteria for approved shellfish-growing waters do not insure that the water or oysters harvested from such waters are free from pathogenic microorganisms."[14]

An editorial in the *Archives of Environmental Health* for May, 1966, entitled "The Shellfish Problem," said: "The ability of shellfish to act as vectors of human disease has been well documented. Early in this century the consumption of shellfish harvested from polluted water was shown to be the cause of typhoid fever outbreaks, and in the past ten years outbreaks of infectious hepatitis in this country and abroad have been associated with the consumption of oysters and clams.

"This epidemiological evidence has been supported by a number of laboratory studies which have shown that oysters and clams will not only take up a significant amount of virus or bacteria from contaminated water, but will also concentrate these organisms to a level up to 60 times that of the surrounding water. Naturally occurring contamination with certain types of echovirus and coxsackie virus has been demonstrated in oysters taken from water frankly polluted with human wastes.

"Paradoxically, shellfish grow best in areas most likely to be contaminated with human waste. . . . "[15]

California mussels become poisonous because they consume the tiny organisms that cause red tide. Poisoning in human beings from eating toxic shellfish has occurred in many places throughout the world.[16]

Current inspection procedures for fish and for water do not adequately protect us. Take the case of the 248,000 tins of salmon that had to be recalled from an Alaskan firm. The fish was distributed nationwide under thirty-nine brand names.[17]

Efforts to pass a fish inspection bill in 1971 may be successful, not because of the horror stories told by the FDA inspectors described in this chapter, but because of the curiosity of a chemist at the State University of New York in Binghamton—Bruce McDuffie, Ph.D. Dr. McDuffie, after a chance remark made by a graduate student, decided to test tuna fish for mercury. He found the fish contained more mercury than the .5 parts per million allowed by the FDA.[18]

Mercury, sometimes called "quicksilver," is a brain, liver, and kidney poison when absorbed in sufficient amounts. How much is enough to cause damage? This is still a matter of controversy.

The FDA instituted a nationwide testing program after Dr. McDuffie's finding and discovered that tuna and swordfish did have too much mercury. More than a million cans of tuna were recalled and, in 1971, the FDA banned swordfish from the market because of mercury contamination.[19]

A paper presented at the American Chemical Society meeting March 31, 1971, in Los Angeles, revealed that a

five-year study of commercially important fish in western
and central Lake Erie showed mercury in excess of federal
guidelines. The survey also revealed that plankton and
algae at the mouth of the Buffalo River contain 70 to 80
parts per million of mercury. Fish feed on the plankton
and algae.[20]

In 33 of the 50 states, mercury contamination of water
has been found in recent months. Livers of the Alaskan
fur seal were found to contain up to 116 times the
permissible level of mercury, forcing the FDA to withdraw 25,000 iron supplement pills containing seal liver
extract.[21]

Tests by the Center for Disease Control in Atlanta,
Georgia, have shown ten times the safe level of mercury
in the hair of Pribilof Island Aleuts, who eat seal meat
and liver.[22]

In 1971, before the Senate fish hearings, a scientist
reported the first human case in the United States of
mercury poisoning from tuna fish eaten by a Long Island
woman, although there was some controversy over the
validity of the case.[23]

In recent months, lead,[24] arsenic,[25] DDT,[26] cadmium,[27] and cancer-causing nitrosamines[28] have been
reported in fish. Certainly, the fishermen are not responsible for such contamination. These products are pollutants
dumped in the waters by manufacturers of plastics, metals,
chlorine, and sodium plants. Mercury flows into the
streams from the manufacturer and from the compounds
on land used to prevent fungus on 80 percent of seed crop
in the United States.

However, the seafood industry must realize at last that
protection and inspection of fish products is in the best
interest of everyone. A survey of eleven major cities,
taken by the *New York Times* after publicity about mercury in fish, showed that American sales of canned tuna
were off 25 to 50 percent, and sales of fresh swordfish,
prior to being banned, were off 10 to 18 percent.[29]

If nothing is done to protect fish in the waters and to
inspect fish destined for the American dinner table, there
will be fewer and fewer fish meals in the future.

HOW CAN YOU TELL
WHETHER A FISH
IS WHOLESOME?

In its frozen state, you can't!
When it is fresh, you should observe the following:

1. Is it of good color?
2. Does it smell? A fishy odor does not belong to fresh fish.
3. Are the eyes clear or clouded?
4. If you press down, does it leave an indentation in the fish? Fresh fish flesh springs up.
5. Is the skin smooth to the touch, and firm?
6. Are the gill sections clear and intact?
7. Is the tail broken?
8. Are the scales loose and lacking in sheen?

5 NOTES

1. Robert C. Stanfill, District Director, Food and Drug Administration, "A Case Study of a Chemical Food Poisoning Involving Fish," presented at the 44th annual meeting of the Central Atlantic States Association of FDA officials, Atlantic City, May 26, 1960.
2. Kenneth J. Kirk, Assistant Commissioner of Compliance, FDA, before the Consumer Subcommittee, Committee on Commerce, United States Senate, July 20–21, 1967.
3. James Brooker, Director of the Division of Fisheries Inspection, personal communication with the author, February, 1971, and Hearings before the Consumer Subcommittee of the Committee on Commerce. U.S. Senate, 1, 2, and 14, 1969, Washington, D.C.
4. U.S. Senate Hearings, 1969, *op. cit.*
4a. *Ibid.*
4b. "Frozen Fish Sticks," *Consumer Reports*, September, 1970, pp. 545–547.

4c. Senate Hearings, *op. cit.*
5. Philip R. Lee, M.D., Assistant Secretary for Health and Scientific Affairs, Department of Health, Education, and Welfare, before the Consumer Subcommittee, Committee on Commerce, U.S. Senate, April 23, 1968.
6. Eugene J. Gangarosa, M.D., *et al.*, "Epidemic of Febrile Gastroenteritis Due to Salmonella Java Traced to Smoked White Fish," *American Journal of Public Health*, January, 1968, Vol. 58, No. 1.
7. Senate Hearings, *op. cit.*
8. James Ackert, President of the Atlantic Fishermen's Union, testimony before the Senate Committee, April, 1968.
9. Leroy Houser, *loc. cit.*
10. National Geographic Society news bulletin, March 13, 1968.
11. George A. Michael, Director of the Division of Food and Drugs, Commonwealth of Massachusetts, Senate hearings, April 24, 1968.
12. Warren G. Magnuson, Senator, Washington, Chairman of the Committee on Commerce Hearings.
13. Werner A. Janssen and Caldwell D. Meyers, *Science*, February 2, 1968, Vol. 159.
14. L. W. Slanetz, Ph.D.; Clara Bartley, Ph.D.; and K. W. Stanley, B.S., "Coliforms, Fecal Streptococci and Salmonella in Seawater and Shellfish," presented before the Laboratory Section of the American Public Health Association at its 95th annual meeting, in Miami Beach, Florida, October 26, 1967.
15. May, 1966, Vol. 12.
16. Edward J. Schantz, U.S. Army Biological Laboratory, Fort Detrick, Md., *Biochemistry* 5:1191, 1966.
17. *Consumer Reports*, April, 1967.
    Lowell Wakefield, president of Wakefield Fisheries, Port Wakefield, Alaska, Senate hearings, 1968.
18. Nancy Hicks, "Mercury Detective," *New York Times*, May 21, 1971, p. 31.
19. United Press International report, December 16, 1970, Washington, D.C.
20. FDA announcement, December 27, 1970.
21. Dr. K. K. S. Pillay, paper presented before American Chemical Society, March 31, 1971, Los Angeles, California.
22. "Toxic Mercury Level Is Sound in Seals," *New York Times*, October 30, 1970.
23. Hearings Before the Consumer Subcommittee, U.S. Senate Committee on Commerce, May 20, 1971, Washington D.C.
24. Peter Bernstein, "Tainted Meal: U.S. Accidentally Finds Lead Traces in Fish," *Newark Star-Ledger*, Washington, Bureau, April 7, 1971.

25. "Fish Contain Arsenic, Selenium," *Science News,* March 20, 1971, Vol. 99, p. 198.
26. Associated Press report, "Four Tons of Salt Water Fish Seized as DDT Contaminated," January 1, 1971.
27. Richard D. Lyons, "3 Fish Caught Near a Battery Factory on the Hudson Contain up to 1,000 Times Normal Cadium," *New York Times,* June 13, 1971, p. 23.
28. Thomas Fazio, *et al., Journal of Agricultural and Food Chemicals,* March 3, 1971.
29. David A. Andelman, "Disclosure of Mercury in Fish Is Hurting Sea Food Industry, *New York Times,* January 10, 1971.

6

Salmonella Sam and the Ptomaine Picnic

Salmonella organisms—named for the American veterinarian Dr. D. E. Salmon—occur in the intestinal tract and tissues of infected humans and animals. Many of the more than 1,200 different types can cause food poisoning.[1] They enter the food supply through meats or animal products from infected animals or from contamination by an infected animal or person. The common symptoms are diarrhea, abdominal cramps, fever, and sometimes vomiting. Infections range from moderate, with recovery in three to four days, to fatal.

In 1970, salmonella accounted for 4,747 reported cases in forty-eight outbreaks of food poisoning. But health officials say the true incidence is probably closer to two million cases a year, with a large percentage of those cases diagnosed as "flu." It is estimated that 1 percent of the population of the United States becomes infected each year.

Salmonella can be deadly. A New Jersey housewife prepared a traditional Thanksgiving dinner in 1968. She purchased a 23-pound frozen turkey, thawed it from the freezer at the bottom of the refrigerator during the morning prior to Thanksgiving Day, and cooked it at 300 degrees F. for 7 hours on the holiday. It was consumed immediately afterward. Three to seventeen hours after the meal, 16 persons were hospitalized. The housewife and her 17-year-old son died of salmonella poisoning. The family's pet dog also succumbed after eating scraps from the table.

In July, 1970, 25 patients in a Baltimore nursing home also died of salmonella poisoning.

Twenty-seven persons were hospitalized with acute digestive upsets in Spokane, Washington. The diagnosis was "salmonella."

Doctors studying the cases knew that the persons involved probably contracted the illness within three weeks of one another—but where did they get infected, and how? There seemed to be no logical pattern. The age distribution of the patients ranged from babies to persons in their seventies. They lived in widely scattered areas within a radius of fifteen miles. There was no recognizable genetic or social relationship among the patients. The only apparent similarity was that the two youngest victims were siblings. Many of the persons involved were not much help because—as is often the case with food poisoning—so much time had elapsed between ingestion of the contaminated food and the diagnosis of the symptoms.

Persistent questioning and complete medical histories did pay off. Twelve patients had something in common after all. Six had eaten chocolate meringue pie, and six had eaten banana meringue pie from twelve to forty-four hours before becoming ill. The pie had been consumed in eleven establishments; only two of the twelve persons had eaten in a common restaurant.

One of the patients, a college student, remembered that there had been an outbreak of illness among residents of dormitories at his college at the same time that he had been hospitalized. Consultation with the college physician and nurse revealed that forty-two persons had been admitted to the college infirmary with upset stomachs on October 1, the same day that the hospitalized student had become ill.

Further detective work on the part of the health officials investigating the case showed that the college cafeteria had received eighty chocolate meringue pies at noon on September 29. The pies had been baked early that morning and then served between 5:00 P.M. and 6:00 P.M., without prior refrigeration.

Two of the hospitalized patients who first drew attention to the outbreak reported eating banana meringue pie at the same hotel dining room during a church luncheon on September 24. Sixty-five of the 180 women attending the luncheon were reached by telephone in several cities. Thirty reported having experienced the salmonella symptoms. All of them had eaten the banana pie.

When the facts were in, the trail led to one bakery.

Swabs were taken from thirty-six surfaces around the bakery, and sixty-three samples of nineteen ingredients in the pies were cultured in the laboratory. All the cultures proved to be negative for salmonellas. There had been no recent illness of any kind, and no diarrheal illness within the past year in any of the fourteen employees, according to interviews.

Only two ingredients were suspect as far as salmonellas were concerned. One of these—spray-dried egg albumen—used in the pie filling and produced in Kansas City, Missouri, was cultured. The cultures of fifteen samples were found negative. The other suspected ingredients, frozen egg white, produced in a Spokane plant and supplied regularly to the bakery for meringue production, proved a stumbling block.

There was no egg white left that had been used in the suspected pies. The eggs were obtained by the creamery for breaking from a large egg-production area spanning twenty states. The plant manager was unwilling to provide samples for laboratory testing or detailed information regarding the purchase of eggs, production of frozen and dried eggs, or the sale of these products.

The next fall, at the same college previously involved in the salmonella outbreak, sixty-four hapless students again came down with gastrointestinal symptoms of infection. Public health officials asked two nearby hospital laboratories to culture samples from patients hospitalized with gastrointestinal illness. Again, salmonella was reported in a number of cases, and once more the patients had eaten meringue pie at one of several local restaurants from ten to thirty-two hours before onset of their symptoms. The pies were from the Spokane bakery.

This time, a decision was made to investigate the egg-breaking plant. On October 4, 1963, the plant's records were seized by court order, and the entire stock of frozen and dried egg whites, yolks, and whole eggs were placed under seizure after being identified in several cities.

Suppliers of the eggs to the breaking plant were still numerous and widely scattered. Yet, representatives of the Food and Drug Administration and the Washington State Department of Agriculture found that the eggs used for breaking were, in general, those that could not be graded. They consisted largely of checked, cracked, and dirty eggs. Samples examined in the laboratory showed that approximately 20 percent of the eggs were unfit for human consumption according to FDA and the Washington State Health Department standards.

Furthermore, it was found that the production of bulk frozen egg whites, yolks, and whole eggs in this plant involved hand-breaking into cups and separation of whites and yolks by employee's fingers! The eggs were then sold to bakers, candymakers, and other users of raw, frozen liquid eggs.

Of the 1,131 specimens tested, salmonella was isolated from 20.8 percent of the cans of frozen eggs and 30 percent of 81 lots.

The Spokane case was reported by Ernest A. Ager and others in a paper entitled "Two Outbreaks of Egg-Borne Salmonellosis and Implications for Their Prevention" in the February 6, 1967, issue of the *Journal of the American Medical Association.*[2]

Eggs are frequently the cause of salmonella infection, and have been for a long time. In fact, we Americans almost finished off our British allies during World War II with spray-dried eggs. We gave them twenty-two types of salmonella they had never had before, and when Churchill referred to "blood, sweat and tears," he may have meant more than we thought he did.[3]

Egg powder as a constituent of cake mixes has been frequently implicated in salmonella outbreaks. When the Canadians tested samples of cake mixes, 54 percent of them yielded salmonellas. (I stopped letting the children lick the cake-mix bowl after I discovered that!)

Dr. Agar and his associates feel that one of the most threatening outbreaks of salmonellosis ever studied in this country occurred in the spring and summer of 1963 in at least fifty-three hospitals in thirteen states. More than one thousand individuals—including both patients and employees—became infected with *S. derby,* a type of salmonella, either directly from the ingestion of raw or undercooked eggs or as the result of contact with individuals previously infected. It was estimated that eggs from as many as twelve thousand farms, after being handled by processors, wholesalers and distributors, could theoretically have reached any fifteen of the hospitals in three states. With persistent detective work, it was determined that only a limited number of farms in two localities were involved.

Another spectacular outbreak occurred among 1,850

mental patients. One hundred and four became ill after drinking raw eggnog.

But no matter how prevalent, egg-borne salmonellosis is just one fraction of the problem.

In a case reported in the *New England Journal of Medicine,* Dr. David J. Lang and his group described how they were confronted with twenty-one cases of infection of salmonellosis in a short period of time. In each instance, the offending organism was detected as *Salmonella cubana.* Fearing "an explosive common-source outbreak," admissions to the pediatric service were halted.

All food in the kitchen and all personnel were tested for salmonella. None proved to be the carrier. Half of those affected were children, including tiny infants. The remainder were elderly. After weeks of searching, no common denominator was found. Then someone suddenly realized that all the patients affected had had tests for gastrointestinal function.

Among the diagnostic solutions used to test the digestive organs was carmine. With this lead it didn't take long to trace the mysterious outbreak to this red dye made from dried, ground, female scale insects and larvae found in Central America and Mexico. These insects sometimes harbor salmonellas.

One of the patients—a nine-month-old child with congenital kidney disease—died after acquiring an overwhelming salmonella meningitis and blood poisoning.[4]

A report made in April, 1968, in Boston to the annual American College of Physicians meeting, by doctors from Memorial Hospital in New York, noted that 12.5 percent of the cancer patients in that hospital who caught salmonella infections died. "Salmonellosis in such a population is a most serious complication," they noted.[5]

The possibility that salmonella may cause the death of unborn babies has also been raised. Dr. Daniel Widelock, Deputy Director of the Bureau of Laboratories, New York City Department of Health, reported that after injections with killed *S. typhimurium salmonellae,* pregnant mice experienced hypoglycemia and an increased sensitivity to histamine. When both factors were present, he said, fetal wastage occurred. Increased malformations were also noted.[6]

Former Food and Drug Administration Commissioner Dr. James Goddard said: "Salmonella has become a major public health problem which is continually increasing in complexity. Despite new sanitary procedures for manufacturing, handling, storing, preparing and serving food, the incidence of the disease is rising throughout the world.

"In addition, formerly predictable epidemiological patterns appear to be changing. For the past thirty years, *S. typhimurium* has been the most common isolate. It was often the unwelcome guest at 'church socials.'

"Now, *S. Heidelberg*, considered rare prior to 1950, is the runner up to *typhimurium*.

"Evidence is accumulating that warns us that salmonella serotypes are mutating and finding a wider range of hospitable hosts among the various animal species," he said in a 1967 article in *Nutrition Today*.

"Milk and milk products spread salmonella too. Disease transmission through pasteurized milk and milk products is rare. But the experience with the instant non-fat dry milk in 1966 shows the potential.

"Swine and cattle are also prime sources of salmonella infection," Dr. Goddard said. "Meat chosen at random in retail markets has shown a high incidence of contamination. This is especially true for pork products. A study in Florida revealed that salmonella contamination of fresh sausage and pork ranged from 8 percent in samples from national distributors to 58 percent from local abattoirs."[7]

In a Massachusetts study, 40.4 percent of chickens inspected in a plant were contaminated with salmonella.[8]

Dr. Goddard said animal feed is another source of contamination; and although the Food and Drug Administration has laid down the law about using feed containing salmonella, they have no real way of policing it. The former FDA commissioner said that household pets are also a source of salmonella infection:

"This is especially true when children feed their pets such products as dog candy. It isn't uncommon for children, particularly crawling infants, to eat the candy too. In addition, the infection of a household pet, even if low grade and not acute, may be a source of infection to human members of the household. One study in Florida showed that 15 percent of 1,626 normal household dogs and 12 percent of 73 normal cats were harboring salmonella."[9]

Some other things around the house can also carry salmonella. Cockroaches, which all too frequently contaminate our food in processing plants, warehouses, restaurants, and supermarkets, as well as in the home, have been incriminated in the transmission of the "bug." Studies indicate that if roaches deposit their fecal pellets on food or food utensils, the salmonella organisms will remain alive for many weeks. Like the housefly, the odious cockroach, by its indiscriminate movements from excrement and other filth to our food, mechanically transmits diseased organisms to humans.[10]

Rats and mice are other carriers of salmonella. The *American Journal of Public Health* reported that salmonella would remain viable in rat feces for 148 days. The presence of whole or ground rat or mouse excreta in feeds of laboratory rodents and other animals has caused outbreaks of salmonellosis and other diseases. The salmonella serotype responsible for mouse typhoid is also responsible for food poisoning among humans.

If you don't believe that it is possible for you to get insect- or rodent-contaminated food, all you have to do is read one page of the monthly *FDA Papers*. Take page 30, in the February, 1967, issue:

● Eggs, frozen whole, Roxbury, Massachusetts: contain salmonella microorganisms; decomposed.[11]

● Cashew nuts, shelled, Portland, Oregon: moldy, decomposed; insect contaminated.[12]

● Flour, Mobile, Alabama: held under insanitary conditions; rodent contaminated.[13]

● Fruitcake, Salem, Virginia: prepared and packed under insanitary conditions; insect and rodent contaminated.[14]

● Lima beans, pinto beans, Alma, Arkansas: held under insanitary conditions; insect contaminated.[15]

These are just a few of the many items on one page in one magazine. Because inspectors do no more than spot-check, you can be sure that such contaminated food does get to your dinner table.

An epidemiological investigation of the gastroenteritis outbreak aboard a Navy ship shows how easy it is to spread salmonella and how vulnerable a health inspector may be. The outbreak occurred in two phases. The first had a symptom attack rate of 2.6; the second had a rate of 4.4 percent among those eating the suspect meal. All the men suffering from nausea and diarrhea had eaten the suspect food. Culture studies disclosed that at least 10 percent of the personnel carried the salmonella infection without symptoms.

In the first outbreak, roast loin of pork was the suspected food; in the second, turkey sandwiches prepared and served several days later seemed to be the culprit. Prior to Thanksgiving aboard the ship, frozen turkeys were removed from the freezer and placed on a thawing table. During this process, the roast pork was carved on a cutting board placed on the thawing table. The juices from the thawing turkeys apparently contaminated the cutting board. In the second outbreak, the same cutting board, which evidently had been improperly cleaned, was used to slice the leftover turkey for sandwiches.

Investigators took surface samples following both outbreaks. In the first outbreak, the sample taken from the thawing table yielded a culture of *S. chester salmonella*. In the second incident, one of the investigators, while inspecting the galley, touched only one item: the cutting board. He then smoked a cigarette without first washing his hands. Two days later, he was the only sick member of the investigating team. He yielded a culture of *S. chester*.

Adequate surface disinfection and sanitation could easily have prevented the entire episode.[16]

The ability to stop salmonella and the misery it causes seems almost impossible today. Dr. Goddard believes that one of the most dangerous sources of the disease is "Salmonella Sam." He compares him or her to "Typhoid Mary," who was an innocent carrier of that disease. Though such people are free of symptoms, they contaminate all they touch.

"Unfortunately, there appear to be a great many carriers among those who handle our food," Dr. Goddard said. "It isn't that the food industry has a peculiar attraction for Salmonella Sams. Rather, food handlers have a greater exposure to infection because of the presence of salmonella in the food-processing environment. In fact, in some institutions, salmonellosis is regarded as an occupational hazard, and workers suffering infection are compensated for time lost from the job."

BOTULISM

"Why don't you come over for lunch?" Helen Brown asked her friend and neighbor Barbara Mason. "I'll make some tuna-fish sandwiches and we can chat."

While the two women talked, Mrs. Brown took some salad dressing she had made and mixed it with the tuna fish.

"That fish smells kind of funny," Barbara said, picking up the can and sniffing it. "Do you smell anything?"

"No, it smells all right to me. Let me taste some," Helen replied. "It tastes all right!"

She finished making the sandwiches, and then decided to open a can of soup because it was such a chilly day. They were almost through with lunch when Mrs. Brown's mother, Mrs. Konners, dropped in.

"Hi, Mother," Helen said. "Why don't you have some lunch with us?"

"I'm not very hungry," Mrs. Konners replied, "but I'll take a taste of that tuna fish. I want to see if you made it as good as I make it."

The three chatted for a while until it was time for the children to return from school.

That night at dinner—about 6:00 P.M.—Helen Brown complained to her husband that her vision was blurred. She kept taking off her glasses and putting them on again. Without finishing the dishes after supper, she went to bed.

"It's hard to breathe," she told her husband, "and my throat feels tight."

She and her husband thought she was coming down with the flu. They believed she would be all right in the morning.

But at 6:30 A.M., Mr. Brown was awakened by the sounds his wife was making. She was breathing convulsively and could speak only in a whisper. He called the emergency squad, and his wife was taken to the hospital.

But before the ambulance arrived, at 7:30 A.M., Helen Brown was dead.

Her luncheon companion, Barbara Mason, had also begun to feel the same symptoms about dinnertime on the day of the lunch. She complained of feeling dizzy and of blurred vision, and had difficulty in breathing. Later, her movements became somewhat uncoordinated, and she vomited frequently during the night. She and her husband both thought it was the flu until they heard about Helen Brown.

Barbara Mason was hospitalized half an hour after her friend's death. She was given Polyvalent Type A and B Botulinus antitoxin.

Her symptoms continued to progress. On the fourth day after the luncheon, she was given Type E antitoxin. She did not improve. At 5:00 P.M., five days after the fateful lunch, Barbara Mason joined her friend in death.

Helen Brown's mother, Mrs. Konners, did not suffer the symptoms until about twenty-four hours after eating a small portion of tuna salad. She began vomiting, and also complained of a sore throat and some difficulty with her vision. She was hospitalized and given 10,000 units of A and B Polyvalent Botulinus antitoxin. Her recovery was relatively rapid and she was released from the hospital three days later.

Laboratory tests confirmed that the illness that struck the three women was botulism.

Samples from Mrs. Brown's garbage can, including some of the salad dressing, the soup can, and the tuna can, were given to the public health laboratory. Even the can opener was brought in. A pure culture of botulinus was isolated from the tuna can.

The health departments of Michigan and California, of Wayne County, Michigan, and the Detroit district of the Federal Food and Drug Administration immediately began tracking down cans from the same lot of tuna fish. The California State Department of Health closed down the cannery that packed the Dagim Tashorim Brand Tuna. The tuna fish had been imported frozen from Japan, and canned in the California plant.

On the same day the plant was closed down, the New York City Health Department picked up a large number of cans in that city. In all, the health officials retrieved

1,200 cans. Tests showed that at least twenty-one contained the deadly botulinus.

Ralph W. Johnston, John Feldman, and Rosemary Sullivan, bacteriologists with the FDA in Detroit, reported the case in Public Health Reports in July, 1963. They said the cans showed defective seams, revealing that the contamination had taken place after sterilization.[17]

The toxin responsible for botulism is the most powerful and deadly poison known to man. Cobra venom, curare, and arsenic are mild by comparison. Controlled experiments at the University of Michigan have shown that 1/100,000,000,000 part of a gram of pure botulism toxin is enough to kill. It is so potent, it has long been recognized as a prospective weapon for bacteriological warfare.[18]

Botulism was first recorded as a disease in 1735, in Germany. The poisoning at that time appeared to be associated with sausage, and the disease was named "botulinus" from the Latin word for sausage, *botulus*. Not until 1895 was the organism discovered and described as occurring in ham and pork. It was called Type A. Thirty years later, a second type, named B, was found in California soil. Since then, it has been found in soils all over the world. Types A and B are both often associated with vegetables and fruit.

In subsequent years, two more types were discovered, C and D. Type C was found occurring in birds, and has, periodically, created havoc in the population of wild ducks. According to Professor Elizabeth McCoy, of the University of Wisconsin, no human case of type C has yet been authentically reported.

Type D was found associated with disease of grazing animals—horses, cattle, sheep. It has been called "forage poisoning" because animals pick it up by feeding on decomposed vegetation that often contains the growing bacteria. This type, like C, also appears incapable of affecting human beings.

The last type found to date, Type E, was discovered in saltwater fish. Poisoning from it occurs most frequently in countries where much preserved fish is consumed. It wasn't until 1951 that scientists discovered that Type E could also come from freshwater fish.[19]

The victims of botulism usually get it by eating improperly canned foods—meat, fish, or nonacid vegetables.

Once in the body, the toxin is absorbed sluggishly by the intestines. But when it gets into the bloodstream, the consequences are swift. The poison sets up "roadblocks" between nerves and muscles, causing paralysis. Breathing muscles are usually the first to suffer, then the heart muscle.

The largest outbreak of botulism ever recorded occurred in Russia in 1933. Some 230 cases, with 94 deaths, resulted from Type-A toxin contained in stuffed eggplant.

The botulism caused by tuna fish at the beginning of this chapter was the first incident in the United States arising from commercially canned foods in nearly forty years. One exception may be a reported incident, caused by commercially canned mushroom sauce, which occurred in 1941.

In 1963, the same year as the deaths from poisonous tuna fish, nineteen more cases of botulism were reported as being caused by commercially smoked fish. Seven persons died. In the same year, home-processed foods were responsible for twenty-two cases of botulism, including five deaths.[20]

The United States canning industry thus far has maintained a good record for safety. However, on June 30, 1971, a New York banker and his wife shared a can of Bon Vivant vichyssoise soup. The man was dead within a few hours after the meal. His wife was paralyzed. Their blood tests showed botulism poisoning, and so did the remaining soup in the can.

The FDA sent out an immediate alert. Newspapers, TV and radio programs carried the message. The vichyssoise the couple had eaten bore the lot number V-141/USA-71. There were 6,444 cans from the batch that possibly contained the toxin. Complicating the recall, however, was the fact that Bon Vivant produced vichyssoise under 22 brands. Hopefully, all the cans will be retrieved.[21]

But no one knows how many cases of botulism there really are each year. There were seven reported outbreaks in 1970.[21a]

An editorial in the *Journal of the American Medical Association*, November 1, 1965, pointed out that because of the relative rarity of botulism, "the diagnosis could not be in the front of the physician's mind and botulism could be overlooked."

"During publicized epidemics, every news reader will know as much about botulism as the physician," the editorial said. "How many physicians know that they can obtain types A and B antitoxin from ready commercial sources and Type E from the Public Health Service Communicable Disease Center in Atlanta, which began to stockpile significant quantities more than a year ago?"[22a]

How well are we being protected from botulism?

Wesley E. Gilbertson, of the Division of Environmental Engineering and Food Protection, United States Department of Health, Education, and Welfare, testifying before a congressional committee, said of the plant where smoked fish had caused death from botulism:

"We were astounded to see . . . that the finished products very often were not separated in the processing rooms from the raw materials coming in. So instead of having the usual type of food operation, one was looking at something that really in a sense lacked the technology we think the food industry should have." He indicated also that the machines used to smoke the fish lacked instruments for measuring heat intensity.

In his testimony, Gilbertson summed up the danger from botulism in America today by saying, "We are on the borderline of a hazard."[23]

Lloyd Kempe, professor at the University of Michigan College of Engineering, has been working on the problem of botulism for more than ten years. He and research bacteriologist John T. Graiskoski helped identify the strain that killed the women in the case at the beginning of the chapter.

Their studies have shown that the bacteria can produce toxin at normal refrigerator temperatures. The poison, however, can be rendered harmless by cooking at about 150 degrees F. for 15 minutes, although it takes one hour's cooking at 195 degrees to kill the poison-producing botulism bacteria.

Kempe and Graiskoski showed that Type E botulinuses can grow and form toxin at temperatures below 40 degrees. In fact, they said, they can incubate botulinuses for research in laboratory refrigerators.[24]

STAPH——PTOMAINE POISONING

At a company picnic in northern Florida, some 150 employees ate ham, potato salad, and cake. Just as everyone was about to start a baseball game—about ninety minutes after lunch, five persons became extremely nauseated and had severe stomach pains. Within half an hour, fifteen more joined their miserable ranks. Before 10:00 P.M. that night, twenty-three persons required temporary hospital treatment and sixty-three were admitted overnight.

At a church dinner, over half of those who had eaten chicken—about one hundred persons—had become ill within six hours.

Investigation into the company picnic case showed not only that the ham and potato salad had been prepared under insanitary conditions by a caterer but also that a company truck had conveyed the food about one hundred miles the day before without any type of refrigeration. It was placed under refrigeration at the picnic grounds, but by then it was too late.

At the church dinner, the chickens had been cooked the day before and immediately refrigerated. The next morning, the chickens were reheated and cut into quarters with a butcher's meat saw. The chickens were without refrigeration from 10:00 A.M. until 5:00 P.M., when they were again reheated. The hands of the cook who prepared the chicken had numerous small cuts and abrasions.[25]

What had ruined the company picnic and poisoned the church parishioners? The staphylococcus germ. It rivals salmonella as the culprit in food poisoning, and there is much controversy over which causes more cases.

Staph is present on human skin, in the human throat, and in the human nasal areas, as well as in festering human wounds. At mealtimes, skin eruptions and cut fingers are common sources of staph germs. Many foods, once infected, will allow the organisms to grow and produce toxins unless the food is refrigerated. Even if the organism itself has been killed by heat, if sufficient toxin has been produced, the food is still capable of causing the symptoms of staph poisoning. The toxin produced by staph is heat-resistant.

Foodborne illness from staphylococcus is often referred to by laymen as "ptomaine" poisoning. This term has been consistently used since its introduction in 1870. Actually, there is no specific entity or group of substances that properly might be called "ptomaine." The word is derived from the Greek word *ptoma*, which means "carcass" or "dead flesh."[26]

The foods most often connected with staph poisoning are custards, cream-filled pastry, milk, and processed meat and fish. Food handlers with staphylococcal skin infections are the primary source of the illness.[27]

The incubation period is usually two to four hours after ingestion of food containing enterotoxin, or "poison." Then the onset is abrupt, with nausea and vomiting. The attack lasts from three to six hours, which is too long if you've ever had it. Fatalities are rare, and complete recovery is the routine. Diagnosis is usually based on the sudden onset of illness after eating infected food, brevity of symptoms, and rapidity of recovery. Furthermore, the patient is usually one of a number of similarly affected victims.

Considerable progress has been made in the past few years in the procedures for detecting staph poisons in food in the laboratory.[28] In fact, new laboratory techniques saved a cheese manufacturer from financial ruin. Staph toxin was discovered in samples of four million pounds of cheese that had been distributed around the country. Even though the live staph germ itself was not present, the cheese was ordered removed from the market. Faced with bankruptcy, the manufacturer had the cheese tested for the presence of enterotoxin, using the new laboratory techniques. As a result, he was able to release all but approximately 62,000 pounds of the cheese. No report of food poisoning followed the consumption of the released cheese.

Unfortunately, the ability to detect enterotoxins and enterotoxigenicity by seriological methods in the laboratory is limited to only a few laboratories because necessary supplies are not generally available. Furthermore, not all staph poisons have been identified.[29]

In 1970, 4,699 persons were known to have suffered from staph food poisoning, comprising 19.8 percent of all reported foodborne illnesses.[29a]

SHIGELLA

On the evening of April 20, 1965, a group of Army personnel and civilians and their dependents held a banquet in a private hall in San Antonio, Texas, to celebrate the end of the bowling-league season. Some 320 persons heard short speeches and long jokes and everyone had a good time.

Within a week, almost everyone who had attended the banquet was miserably ill. The sickness came on suddenly.

They had fever, felt drowsy, lost their appetites, began to vomit, had severe diarrhea, and abdominal pains with distention. Three days after their symptoms began, blood, pus, and mucus began to appear in their stools. Actually, the epidemic was under way for a week before its existence was recognized. The first victims were treated at scattered civilian and military medical facilities. Their cases had been erroneously diagnosed as flu.

Then Lieutenant Colonel Jerome H. Greenberg, M.D., and other health investigators stepped in. They looked over the menu of the food eaten at the banquet. They immediately suspected the potato salad because it is a food often implicated in such illnesses.

The investigators went to the potato-processing plant. They found numerous flies. The bathroom lacked soap and towels. After potatoes were peeled and cooked, they were cooled, unprotected from flies, in front of a fan for several hours. The potatoes were then diced and placed in plastic bags and baskets. During the process, some of the potatoes spilled onto the floor, while others were partly in contact with the floor through the latticework of the baskets. Upon completion of dicing, some four hours after the potatoes were cooked, the bags and baskets were placed in a walk-in refrigerator.

On the day of the banquet, the temperature of the refrigerator had been 52 degrees F., and the internal temperature of the bag of potatoes that had been in the refrigerator approximately three hours was 64 degrees F. No further processing was done until the potatoes were delivered, in an unrefrigerated truck, to the restaurant where the salad was made. Additional periods of time during which the potatoes remained unrefrigerated occurred at the restaurant while the salad was being prepared and later, during delivery to the banquet hall. The durations of the periods could not be determined accurately.

In any event, the unhygienically prepared potato salad was capable of giving three hundred people shigellosis, or bacillary dysentery, which is an acute infection of the bowel caused by organisms of the shigella group and characterized by bloody diarrhea.

The incubation period is one to four days. This is one reason that diagnosis is often erroneous. Suspect food after that length of time is often not available for examination. Another reason is that laboratories often fail to use techniques that will detect a small number of shigellae in the presence of larger numbers of other bacteria.

Shigella infections are usually spread by contaminated water, defective plumbing, or food contaminated by unwashed hands or by flies. Good sanitation, with exclusion of flies and protection of food from contamination, is of prime importance in preventing the infection.

Dr. Greenberg pointed out in an article in *Public Health Reports*, November, 1966, that the disease is endemic throughout a large part of the United States. In epidemic form it has been less common than salmonellosis, but when it occurs, it usually strikes down a majority of those exposed.[30]

There were 1,668 cases reported in 1970, but with shigellosis, as with other foodborne diseases, the true incidence is greatly underreported. Because shigellosis affects many institutionalized patients and has become resistant to antibiotics in many instances, a vaccine against it has been developed. It is now being used in clinical tests to determine its effectiveness.[31]

CLOSTRIDIUM PERFRINGENS

One day, an exclusive girls' school had almost 100 percent illness. From the headmistress to the youngest student, everyone was crowding the bathrooms with diarrhea and colic-like cramps. The suspect meal was roast beef and gravy. The beef and gravy had been prepared the day before and allowed to cool in open trays, without refrigeration, for twenty-two hours.

In 1970, *Clostridium perfringens* accounted for almost 30 percent of all patients and 15 percent of all outbreaks of food poisoning.[31a]

Clostridium perfringens is widely distributed in feces, sewage, and soil. It is so prevalent it is a probable contaminant of nearly all foods. The spores of *C. perfringens* are present in a large percentage of meat products, both raw and cooked.[32] The cooked foods most frequently associated with outbreaks in America are meats, fowl, and gravies that have been cooked, and allowed to cool slowly.

Herbert E. Hall, Ph.D., and Keith H. Lewis, Ph.D., writing in *Health Laboratory Science*, October, 1967, said: "We do not know how many outbreaks involving *C. perfringens* occur each year in the United States. The syndrome is a mild one, and unless an isolated group is involved, medical and epidemiological attention is rarely directed toward its occurrence."[33]

In the American Medical Association publication *Let's Talk About Food*, edited by Philip White, Sc.D., it was pointed out that symptoms from *C. perfringens* last one to one and a half days. "Since the patients recover quickly, and since people learn to tolerate minor episodes of discomfort, thorough surveillance by public health authorities is difficult."[34]

Dr. Hall and Dr. Lewis pointed out that public health laboratories that routinely examine food for isolations of

C. perfringens usually find outbreaks caused by this organism.

"Further, we know that over 75 percent of food handlers carry one or more strains of these organisms in their feces, and that recovery of these organisms from the food preparation environment is not difficult and that 37 to 82 percent of raw meats and 4.7 to 37 percent of processed meats contain these organisms. It is, therefore, reasonable to expect that the etiology of many more outbreaks would be determined if all laboratories examined suspect food anaerobically."[35]

ESCHERICHIA COLI

As Christmas lights were being hung along the streets and while carols poured from loudspeakers in December, 1965, babies began to sicken and die in Newark, New Jersey. The infants had severe diarrhea, a condition that does kill millions of infants in underdeveloped countries around the world every year. But what had caused it in New Jersey's largest city? What did the babies have in common? They came from families of all economic levels and from all ethnic groups.

By the time officials realized it was an epidemic, sickness had struck scores of babies. Public health officials who began investigating the case found that there was one thing 30 percent of the babies did have in common. They had been hospitalized in the recent past, but for a variety of conditions. Without knowing what was causing the sickness, doctors tried desperately to save the infants. They used a wide spectrum of antibiotics, but none seemed of much use.

By the time the epidemic had ended, four hundred infants had been seriously affected and twenty-eight had died. Of those that succumbed, almost all had been in the 30 percent previously hospitalized for other conditions.

Laboratories finally succeeded in isolating the organism that had caused the diarrhea. It was *Escherichia coli*.[36]

Little is really known about the "bug." It is known that *E. coli* is normally present in the intestines. Abroad, it has been reported as a cause of both food and waterborne illness in adults, and it is associated particularly with infant diarrhea in this country. The symptoms range from those of a mild disease to a severe illness with high fever, acute diarrhea, and prostration. In addition to the overt illness caused by ingesting EEC, both the acutely ill and those displaying no symptoms become carriers for variable periods of time, and are potentially dangerous to small infants.[37]

Although they are not sure, public health officials believe that the Newark infants picked up the disease in hospitals, either from food or personnel, during their stay for treatment of other conditions; then they spread the disease to other infants in the community.

There is no doubt that *E. coli* is contaminating some of our food. In 1970, FDA inspectors discovered it in:

- Macaroni and cheese, Carrollton, Missouri.[38]
- Pecan pieces, Honolulu, Hawaii.[39]
- Frozen, breaded onion rings, Macomb, Illinois.[40]

Approximately 20,000 pounds of the prepared cake mixes, valued at $3,000 were destroyed voluntarily because the products were contaminated with *E. coli*.

But how much of our food is contaminated with the organism? How many of our illnesses are caused by *E. coli?* Although proper laboratory tests could tell us, no one knows. *E. coli* is most commonly found in food served in public places, according to the Center for Disease Control in Atlanta.

FOODBORNE ILLNESSES
FROM VIRUSES

A kitchen worker in a hospital in St. Louis, Missouri, went about her duties, although she was worried about her husband. He didn't complain much, but he acted as if he didn't feel too well. The date was July 2. As she mixed the frozen orange-juice concentrate with water for the next morning's breakfast, she thought to herself that she was glad she was well and able to hold down a job to supplement the family's income. She placed the orange juice in the refrigerator, and went home.

The next morning, interns, resident physicians, nurses, and cafeteria employees filed in for breakfast. Most of them, as is the custom in America, took orange juice.

Between July 21 and August 12, fourteen of the orange-juice drinkers came down with hepatitis. This was in spite of the fact that the orange juice had a pH (acidity-alkalinity content) in the range of 3.5 to 4.0 and had been in the refrigerator at least overnight. At the same time the hospital personnel came down with hepatitis, so did the kitchen worker's husband. The public health officials believe that he had the virus and that his wife carried it without symptoms.[41] For some reason, young people seem to be affected most frequently by hepatitis. The peak season apparently is in the fall. The incubation period for hepatitis is estimated at two to six weeks.[42] The onset of the symptoms is abrupt, with loss of appetite, nausea, fever, and malaise. Tenderness and enlargement of the liver and pain in the right upper quadrant are usually early symptoms. About five days after the onset, jaundice appears, and fever tends to subside. The gastrointestinal symptoms persist for about ten days, and subside with regression of the jaundice. Swollen glands, severe generalized itching, hives, and intermittent diarrhea may occur. Usually, the patient recovers uneventfully after six to eight weeks. Relapses occur in 5 to 15 percent of the cases, and are usually attributed to premature resumption of activity, poor diet, or alcoholism. Ninety percent of the patients recover.

Dean O. Cliver, writing in the October, 1967, issue of *Health Laboratory Science,* said foodborne viruses are usually too small to be seen with a regular microscope.

"Since the laboratory diagnosis of virus disease is relatively new, the greatest emphasis in the epidemiological record has necessarily been placed upon diseases whose clinical pictures were so distinctive as to permit them to be diagnosed on that basis by the attending physician. The first of the clinically distinctive human agents reported to be transmitted by foods was poliomyelitis. Ten outbreaks took place between 1914 and 1949 . . . milk, particularly raw milk, was the predominant vehicle."

He went on to say, "Although there is no reproducible method of cultivating the virus of hepatitis in the laboratory, the clinical picture is frequently highly distinctive, and epidemiologists are coming increasingly to consider food as a possible vehicle for the virus." He added that since the incubation period for hepatitis is approximately four weeks, it is extremely difficult to compile definitive information once an outbreak has been recognized.[43]

In another article, in *Public Health Reports*, February, 1966, Dr. Cliver said that the majority of foods implicated in the outbreaks of hepatitis had been cooked very little: "It is to be noted that contamination which takes place almost immediately before consumption is the least likely to be counteracted by inactivation of the virus as a result of cooking or other treatment of the food. As a result, the infected human who works either as a food handler or a kitchen worker in final preparation constitutes a potentially quite significant source of virus in foods."

As with the spread of bacteria-borne food poisoning, good sanitation is one step in prevention. As far as the hepatitis virus is concerned, public health officials agree that the virus is shed by infected individuals in the feces, so that considerable mishandling of food must occur before a foodborne outbreak.

A demonstration of how poor the reporting of foodborne illness really is can be determined in the following figures. In 1970, there were only 107 cases of viral hepatitis reported to the Center for Disease Control in Atlanta, Georgia. In May alone, in 1971, New Jersey recorded 255 cases of infectious hepatitis; in 1970, 158 were reported.[43a 43b]

FOODBORNE PARASITIC DISEASES

Corner grocery stores are fast being replaced by the impersonal supermarket. But some small stores still exist in ethnic neighborhoods. Practically everyone went to one such market to get "homemade" pork sausages. "They taste just like the ones in the old country," people would say. Even young children who had never been out of the boundaries of the neighborhood liked them.

One spring, eleven people in the neighborhood, ranging from a four-year-old to an eighty-year-old, came down with high fever, muscle pain, stomach cramps, chills, and general weakness. They were members of seven families, but the cause of their symptoms wasn't hard to track down. All had eaten raw smoked sausage prepared from the same hog at the corner grocer's. The hog was infected with *Trichinella spiralis,* a small parasite.[44]

Trichinella spiralis requires two hosts, animal and man. Trichinosis results when humans eat insufficiently cooked flesh of infected animals. Pork is the most common source of the trichina-infected meat.[45] The larvae originally held in the animal are released from their "space capsule" in the small intestines of the person who consumes the meat. Their parasites mature within a few hours and mate within two or three days. The male worm of *T. spiralis* is eventually passed out of the human body. The females, on the other hand, burrow deep into the intestinal wall, and there they discharge many larvae over a six-week period. These offspring enter the bloodstream, and are carried to the heart, where they are then pumped to all parts of the body. Only those that lodge in muscle tissue are capable of further development. Within a month, they are encapsulated by fibrous tissue, and the cycle is complete. They may live indefinitely in these cysts without causing any significant symptoms.[46]

However, persons who have eaten a large number of the worms in their food usually develop such symptoms as swelling of the upper eyelids, muscular tenderness, headaches, diarrhea, and fever. They may have an eye hemorrhage, eye pain, and shy away from light. If the worms affect the heart, respiratory system, or the nerves, severe symptoms and even death may result. But permanent disability rarely ensues, although symptoms may persist for several months.

Though it is hard to believe, it is conservatively estimated that 25 to 50 million Americans carry trichina larvae in their muscles and internal organs. Every American is estimated to consume at least three servings of infested pork annually. Up to 10 percent of the pork sausage in large city markets has been found to be infected, according to Howard Earl in the October, 1965, issue of the American Medical Association's publication *Today's Health*.[47]

Between two and three hundred human cases of trichinosis are reported each year in the United States in spite of the fact most states do not make such reporting compulsory. The fatality rate from trichinosis is listed at about 5 percent, but in epidemics it has been known to reach more than 30 percent.[48]

The prevention of trichinosis should be simple. Thorough cooking of all fresh pork and the feeding of only cooked garbage to hogs would go far toward reducing it. About one-fourth of 1 percent of the grain fed to swine, and considerably more of the raw garbage, are estimated to be infected. Although the law states that garbage should be cooked, it is difficult to enforce, and infected animals are still on the market.[49]

TOXOPLASMOSIS

In 1965, the National Academy of Sciences urged further research in the field of parasitology. Tapeworms, protozoa, and other parasites can cause all sorts of symptoms, from ulcers to weight loss to dysentery.[50]

Disease from the protozoa *Giardia* is present in 7.4 percent of the people in the United States and only 6.9 percent of the people in other parts of the world.[51] Amebic dysentery is endemic throughout the world, affecting 17.6 percent of the population. In the United States, it affects 13.6 percent.[52] All can be spread through unsanitary food. No one, as we have said before, really knows the extent of the parasites and the diseases they cause.

An example of how important research can be may be found in the example of toxoplasmosis. Since the 1930's, it has been known that *Toxoplasma* can cause significant disease in man. It was then shown by physicians at Columbia University that this parasite could be transmitted from mother to unborn child. But in the intervening years, it was never revealed:

● How the mother becomes infected.
● What the true incidence of congenital toxoplasmosis is.
● How an early diagnosis can be made.
● How preventive measures can be taken.
● How widespread mental retardation, compromised vision, epilepsy, hydrocephalus, and so forth, are because of toxoplasmosis.[53]

There is no doubt that toxoplasmosis cripples unborn children. One study showed that of 150 infants with proved congenital toxoplasmosis, approximately one-third showed signs and symptoms of an acute infectious process with splenomegaly, jaundice, enlarged liver, anemia, retinal damage, and abnormal cerebrospinal fluid.

How many babies have been born with damaged brains and blind eyes because of toxoplasma gondii?

Dr. Jack S. Remington, Chief of the Division of Allergy, Immunology, and Infectious Diseases, Palo Alto Medical Research Foundation, reported in 1967 that detailed recent studies supported the view that ingestion of infected meat or water may be the mode of transmission. He noted that studies of animals chronically infected with toxoplasma had shown that the cysts obtained from muscle were resistant to gastric acidity and digestive juices. The findings thus suggested that undercooked meat played a role in spreading the disease.[54] In studies by his own group of fifty samples each of beef loin, lamb and pork loin, obtained from local stores, Dr. Remington said toxoplasma was found in 4 percent of the lamb samples, 32 percent of the pork samples, and none of the beef samples.

Further support to the meatborne theory was the case study of a French hospital for tubercular children. At the time of admission, 641 children showed no toxoplasma antibodies; but after admission, 204 did. The risk of infection per month of hospital stay was approximately five times as high as for the general population. When the usual ration of undercooked meat was increased, the number of infections showed a significant increase, and the risk almost doubled.

Dr. Remington said of the French hospital study: "The custom in this hospital of eating raw or very undercooked meat was probably the cause of the abnormally high number of toxoplasma infections. The authors stated that the prevalence of toxoplasma infections in France is higher than in other countries with similar climate and culture. The difference is particularly pronounced among children."[55]

Dr. G. Richard O'Connor, Associate Director of the Francis I. Proctor Foundation for Research in Ophthalmology, is studying the effect of toxoplasmosis on the eyes of newborn infants. "We think there is a possibility that mothers may transmit the disease to their infants. When the disease occurs, we find the one thing many of these mothers have in common is that they have eaten raw meat sometime early in their pregnancy." Mothers, he emphasizes, should not eat raw meat during pregnancy and should also avoid raw eggs.[56]

OTHER FOODBORNE DISEASES

There are many other foodborne diseases. In fact, more than 40 percent of all human diseases are contracted through food.

Typhoid fever, although it crops up once in a while, is not too much of a problem in America today. It can still be a danger in water, milk, and shellfish that is contaminated at the source. Food contaminated by unwashed hands or by flies may still give a person typhoid fever.

Streptococcus, which causes septic sore throat and scarlet fever, may still be obtained from raw milk contaminated at the source. Diphtheria is a remote danger from dishes or silverware contaminated by a carrier.

Brucellosis, or undulant fever, is an infectious disease with an acute high fever but few other signs. Vague body aches, weakness, and sweating may also be associated with it. Brucellosis is too common in the United States today. The three causative microorganisms belong to the genus *Brucella*, of which three species primarily affect animals: *Br. abortus*, cattle and hogs; *Br. suis*, hogs; and *Br. melitensis*, goats. Any of these may be transmitted to man by direct contact with secretions and excretions and by the ingestion of milk or milk products containing viable brucellae. Though brucellosis is a worldwide disease, it is most prevalent in rural areas, and is an occupational disease among meat-packers, veterinarians, farmers, and livestock producers.[57]

The symptoms may simulate a number of human diseases, making diagnosis very difficult. In the heart of Newark, at the United Hospitals, a number of women who had miscarriages were tested. The lesions in their placentas were similar to those seen in the placentas of cattle who had lost their calves. The brucellosis organism was recovered from the women.[58]

NATURAL FOOD POISONS

Dr. Walter MacLinn, Director of the Agricultural Experiment Station at Rutgers University, says: "You must remember that animals and plants were not put on earth for the benefit of man. Everything has just as much right to exist as man. Man is a predator. Some of the things he eats are going to make him sick."[59]

Potatoes, Brazil nuts, and tangerines are among the items that are inedible for some people. Dr. Thomas Golbert, a fellow in medicine at Northwestern University, described thirteen cases of anaphylactic shock due to common foods. Besides potatoes, nuts, and tangerines, there were pinto beans, halibut, rice, shrimp, milk, a cereal mix, and garbanzo (chick-peas).[60] All the reactions were acute and potentially fatal, with skin, respiratory, and blood-pressure involvement. Swelling, hives, or both, occurred in every case. Thirteen of the patients lost consciousness.

In several instances, Dr. Golbert said that he and his colleagues examined the patients in the emergency room. In others, a complete medical and allergic history was done. In eight cases, the cause was incontrovertibly proved: the patients developed anaphylactic shock a second time after ingesting the same food.

In eleven cases, once the cause was proved or was presumptively identified, there were no recurrences. In two cases the cause of the reactions remains uncertain. One patient has had several recurrences after eating foods with a large mold content: cheeses, old breads, certain kinds of sausage—and he has reacted strongly to a test with mold antigen. The last patient is still a mystery, although an investigation of hidden additives may explain why he shows no reaction to extracts of the pure foods during tests.

A. D. Campbell, Chief of the Food Contaminants Branch, Division of the Food Chemistry, Food and Drug Administration, discussed "natural food poisons" in the September, 1967, *FDA Papers:*

"Generally, through a trial-and-error process, man has been able to place plants and animals into safe and unsafe food categories. This trial-and-error process has been useful in eliminating those which produce acute toxicity; the results are usually sudden illness or death. However, the manifestations of chronic toxicity of some of these substances are not readily associated with the source; symptoms may occur after a considerable lapse of time. More sophisticated research studies are usually necessary to find the cause-and-effect relationships for chronic toxicity."

He went on to say that plants may contain toxic substances, such as protein digestion inhibitors, toxic proteins such as those found in castor beans, estrogenic substances, substances containing cyanides, solanine substances such as toxins from potato sprouts, and other toxic materials.

"In some instances, the toxicant is at such a low level that it does not present a health problem when eaten; the estrogenic substances of soybeans are examples. In other instances, the toxicant can be inactivated by heating the foodstuff before it is eaten. For example, the protein digestion inhibitor of soybeans and the cyanide substances in lima beans are inactivated, as toxicants, by heating.

"The toxic portions of some plants are usually removed and discarded. Those found in the skin and sprouts of potatoes which have been exposed to sunlight are thus avoided. Stalks of rhubarb are eaten, and the toxic leaves are discarded.

"Harvesting at the proper stage of maturity is a means of avoiding other plant toxicants. For example, unripened grapefruit contain a toxic substance which is not present in mature fruit. The use of selective breeding by the plant geneticist is another means of eliminating toxic substances from some plants."

Dr. Campbell pointed out that some animals or animal products that are normally suitable as food can contain toxic substances under unusual circumstances:

"Oysters, clams, and other shellfish have been shown to contain a toxic substance known as 'the paralytic shellfish poison' when they grow under adverse conditions. The puffer fish, considered a delicacy by some, contains a highly toxic substance in its skin and sex organs. Toxic animal metabolites (substances produced by the life processes) are known to contaminate some animal products when some toxic substances from molds are eaten by the animals. Detection by analytical means and discarding the contaminated lots may be the only means of avoiding them."

The FDA scientist then launched into an extensive discussion of the mycotoxins that have been with "man since the beginning." He referred to the incident in 1961 when hundreds of English turkeys died from a toxic substance produced by the mold *Aspergillus flavus*. The substance was called "aflatoxin."

"It is known that the aflatoxins are produced by a number of molds in addition to *Aspergillus flavus*," Dr. Campbell said. "These are very potent toxins for some animals, and the sensitivity varies over a considerable range for different species. Rainbow trout are the most sensitive animals that have been found so far."

Dr. Campbell said it is not uncommon for the mycologist to isolate molds from cereal grains and other foodstuffs. He said that American scientists began to work on the mold toxin problems after 1961. As a result of FDA's early efforts, and through the cooperation of industry and the United States Department of Agriculture, contaminated peanuts have been diverted from food channels.

Each lot of peanuts, he said, is examined as it comes from the farm, and the lots likely to be contaminated with aflatoxins are analyzed. If the contamination is found too high, the lot is used only for the production of peanut oil, because the refining process eliminates aflatoxins.

The aflatoxin problem has not been confined to peanuts. Dr. Campbell concluded:

"Much has been learned by the scientific community from the great amount of research which has gone on in many laboratories since the aflatoxins were discovered in 1961. This information has been quickly put to use to protect the health of the people throughout the world. These studies indicate that much is still to be learned about naturally occurring food poisons. Only through concerted research efforts of many laboratories will the basic scientific information be obtained for a sound evaluation of this newly recognized problem."[61]

If aflatoxin is a newly recognized problem, death from mushrooms is a very old one, though mushrooms are far from the only plants carrying deadly poison. Although they kill rapidly, some may kill slowly.

A detailed report on foods that certain substances known to have toxic effects on man under some conditions was issued by the Food Protection Committee of the National Research Council in 1967. The Academy of Sciences release said of the report: "*Toxicants Occurring Naturally in Foods* suggests the need for continuing research in a field that has received comparatively little study and which contains many unknowns."[62]

The report concludes that the primary medical challenge lies in "the question of the long-term chronic toxicity, or lifetime effects, of many of the known as well as the yet unrecognized natural chemical components of our foods.

Among the many natural food chemicals with possible long-term harmful effects are those that can produce goiters, tumors, and cancers. The foods that carry the greatest amount of goitrogenic chemicals include rutabagas, turnips, kale, and cauliflower. These chemicals are also present in wild turnips and in such weeds of the mustard family as lesser swine's-cress, pepper cress, and shepherd's purse. The report said that the chemicals "appear to be transferable to the milk of cows consuming these plants and to be capable of interfering with the thyroid function of humans who drink such milk."

In discussing the chemicals that have produced cancer in laboratory animals and that are found in some humans' foods, the report cautions: "The role played by these carcinogens in the occurrence of certain cancers in man in regions where these agents are found is still unknown. An informed approach to the hazards for these agents and the synthetic chemicals in our environment is necessary...."

Some of the carcinogens considered in the report are ergot, a fungus substance sometimes found on rye; a mold that causes rice to turn yellow and bitter; an aflatoxin.

Stimulants and depressants are found in many plants. Jimsonweed, used frequently in home medicines as a tea, produces hallucinations. It is estimated that four to five grams of the crude leaf or seeds is fatal. Nutmeg, used by some individuals for its narcotic effects, also produces burning abdominal pain, delirium, stupor, low blood pressure, liver damage, shock, or death. The amount of nutmeg found to cause poisoning is inconsistent, but the report cites one study that sets the amount at five grams or more.

The report confirms that caffeine, in the amounts present in coffee, tea, and many cola drinks (a six-ounce bottle of cola contains about one-third the caffeine of a cup of coffee) tends to facilitate mental and muscular effort, and diminishes drowsiness and psychic and motor fatigue. Heavy coffee drinking, along with several other factors, have been associated in a Chicago study with the development of coronary heart disease. The coffee association does not become apparent, physicians pointed out, unless six or more cups a day are consumed.[63] Coffee has also been designated as a food allergen. It has been known to cause migraines, stuffy nose, and upset stomach.[64]

Tea contains, in addition to caffeine, a second, less powerful stimulant—theophylline. Both it and caffeine can be fatal in doses far greater than can be ingested ordinarily in coffee or tea.

Honey made from nectar of the Carolina jessamine, also known as yellow jessamine and trumpet flower, has caused death in man.

Many common foods—such as bananas, pineapples, tomatoes, some cheeses, lemon, and wine—contain large amounts of toxic amino compounds, enough to have disastrous consequences if they were injected into the veins. Among the factors that are believed to be responsible for rendering these amino compounds harmless in foodstuffs is detoxification after consumption through the action of another chemical, monoamine oxidase (MAO). However, severe complications have been reported among patients on tranquilizers following the consumption of pickled herring.

Another type of complication from combinations of foods is mentioned briefly in the report on toxicants: acute illness caused by drinking an alcoholic beverage following the consumption of the edible "inky cap" mushroom. A chemical in the mushroom interferes with normal metabolism of the alcohol.

The list of naturally occurring food poisons and possibilities is endless.

Spinach has been implicated several times. One West German baby died and thirteen others between the ages of two months and ten months developed an anemic blood condition after eating spinach or drinking the water in which it was cooked. Nitrites in the spinach were blamed.[65] A Milwaukee cardiologist warned that heart patients taking anticoagulant drugs should not eat spinach or other leafy vegetables containing vitamin K. Vitamin K aids blood clotting.[66]

An Ohio doctor believes that large amounts of mustard, pepper, and ginger in the diet can produce high blood pressure.[67]

6 NOTES

1. "FDA and Salmonella," *Sanitation Review*, February, 1967.
2. Ernest A. Ager, M.D.; Kenrad E. Nelson, M.D.; Mildred M. Galton, Sc.M.; John R. Boring, III, Ph.D.; Janice R. Jernigan, "Two Outbreaks of Egg-Borne Salmonellosis and Implications for Their Prevention," *Journal of the American Medical Association* (*JAMA*), February 6, 1967, Vol. 199, No. 6, pp. 372–378.

3. *Ibid.*
4. David Lang, M.D., *New England Journal of Medicine,* 276:829–832, 1967.
5. Donald Armstrong, M.D.; Martin Wolfe, M.D.; Donald Louria, F.A.C.P.; Anne Blevin, R.N., Department of Medicine, Memorial Hospital, N.Y., 49th Annual Session, American College of Physicians, April 15, 1968, Boston.
6. Daniel Widelock, Ph.D., Deputy Director, Bureau of Laboratories, New York City Department of Health, "Salmonella Link in Fetal Death?", *JAMA,* December 19, 1967, Vol. 202, No. 12, p. 32.
7. James Goddard, M.D., Commissioner of U.S. Food and Drug Administration, "Incident at Selby Junior High," *Nutrition Today,* September, 1967.
8. Arthur Wilder, D.V.M.; Robert MacCready, M.D., "Isolation of Salmonella from Poultry," *New England Journal of Medicine,* June 30, 1966, Vol. 274, No. 26.
9. Goddard, *op. cit.*
10. "FDA and Salmonella," *Sanitation Review,* February, 1967.
11. *FDA Papers,* April, 1971, p. 38.
12. *Ibid.,* February, 1967, p. 30.
13. *Ibid.*
14. *Ibid.*
15. *Ibid.*
16. Lt. Warren Sanborn, M.S.C., USN, "The Relation of Surface Contamination to the Transmission of Disease," *American Journal of Public Health,* August, 1963, Vol. 52, No. 8.
17. Ralph W. Johnston, M.S.; John Feldman, A.B.; Rosemary Sullivan, B.S., *Public Health Reports,* July, 1963. The names of the victims are fictitious.
18. University of Michigan News Service, October 9, 1968.
19. Dennis Blakeslee, feature story, University of Wisconsin News Service, December 10, 1963.
20. *Ibid.*
21. Miscellaneous press reports, July 3–6, 1971.
21a. Center for Disease Control, Atlanta, Georgia, "Foodborne Outbreaks," *Annual Summary,* 1970.
22. Editorial, *Journal of the American Medical Association,* November 1, 1965, Vol. 194, No. 5.
23. Wesley E. Gilbertson, Division of Environmental Health Engineering, U.S. Department of Health, Education, and Welfare, testimony before House, HEW Appropriations Subcommittee, 1965.
24. University of Michigan News Service, October 9, 1963.
25. Public Health Service Publication 1105, November, 1963.
26. Howard Earl, "Food Poisoning," *Today's Health,* October, 1965.
27. *The Merck Manual,* 11th edition, 1966.

28. E. P. Casman, Ph.D., "Staphylococcal Food Poisoning," *Health Laboratory Science,* October, 1967.
29. *Ibid.*
29a. Center for Disease Control, *op. cit.*
30. Jerome H. Greenberg, M.D., "Common Source Epidemic Shillegosis," *Public Health Reports,* November, 1966.
31. *Journal of the American Medical Association,* "Shigellosis Vaccine Tests Reach Field Trial Stage," January 18, 1971, Vol. 215, No. 3, p. 379.
31a. Center for Disease Control, *op. cit.*
32. *Food Facts from Rutgers,* October–November, 1967.
33. Hall and Lewis, *op. cit.*
34. *Let's Talk About Food,* ed. Philip White, Sc.D., American Medical Association, 1967.
35. Hall and Lewis, *ibid.*
36. News coverage and *Medical World News* report, "PHS Traces Epidemic of Fatal Diarrhea," June 4, 1965.
37. Hall and Lewis, *ibid.*
38. *FDA Papers,* April, 1971, p. 38.
39. *FDA Weekly Recall Report,* October 15, 1970.
40. *Ibid.,* November 25, 1970.
41. Dean O. Cliver, Ph.D., "Implications of Foodborne Infectious Hepatitis," *Public Health Reports,* February, 1966, Vol. 81.
42. *The Merck Manual.*
43. Dean O. Cliver, Ph.D., "Food-Associated Viruses," *Health Laboratory Science,* October, 1967, Vol. 4, No. 4.
43a. Center for Disease Control, *op. cit.*
43b. *Journal of the Medical Society of New Jersey,* May, 1971, Vol. 68, No. 5, p. 445.
44. U.S. Public Health Service Publication 1105, November, 1963.
45. *Let's Talk About Food, loc. cit.*
46. *The Merck Manual.*
47. Howard Earl, "Food Poisoning: The Sneaky Attacker," *Today's Health,* October, 1965.
48. *Ibid.*
49. *The Merck Manual.*
50. C. Cheng, "Parasitological Problems," *Journal of Environmental Health,* December, 1965.
51. *Ibid.*
52. *Ibid.*
53. Richard O'Connor, M.D., Associate Director of the Francis I. Proctor Foundation for Research in Ophthalmology, lecture before the Society for the Prevention of Blindness, Science Writers' Seminar, October, 1967.
54. *Medical Tribune,* January 22, 1968.
55. *Ibid.*
56. O'Connor, *op. cit.*
57. *The Merck Manual.*
58. Samuel A. Goldberg, D.V.M., Ph.D., M.D., "Bacterial and

Parasitic Diseases Common in Animals and Man," *New Jersey Public Health News,* January, 1961.

59. Walter A. MacLinn, Director of the Agricultural Experiment Station, Rutgers University, in a taped interview with author, January, 1968.

60. Thomas Golbert, M.D., American Academy of Allergy meeting, Boston, 1968.

61. A. D. Campbell, Ph.D., "Natural Food Poisons," *FDA Papers,* September, 1967.

62. News from the National Academy of Sciences and the National Research Council, April 19, 1967.

63. Paul Oglesby, M.D., Professor of Medicine, Northwestern University, and Mark Lepper, M.D., Professor of Preventive Medicine, University of Illinois, "Circulation," *Journal of the American Heart Association,* July, 1963.

64. Martin Green, M.D., "Coffee as a Food Allergen," *Journal of the New Jersey Medical Society,* June, 1960, Vol. 57, No. 6.

65. *British Medical Journal,* January 29, 1966, report of Drs. A. Sinios and W. Wodak.

66. Armand J. Quick, M.D., *Science News Letter,* January 25, 1964.

67. Jackson Blair, M.D., *Medical Times,* November 5, 1966.

7

Water:
Unfit for Drinking

A funeral parlor in Atlanta, Georgia, had its hydroaspirator—a device used to withdraw body fluids during embalming—improperly connected to the local drinking-water supply. Several doors down the street, some people were getting something unexpected from their faucets. One man who was questioned by local authorities said, "I thought the water looked rusty, but it wasn't rust."[1]

In New Jersey on a spring night in 1969, a vat of one of the most deadly chemicals known to man, cyanide, sprang a leak. The poison seeped through a drain into the Passaic River running by the plant. The Passaic River provides drinking water for nearly 300,000 persons. Quick action by both plant executives and state and local water officials prevented a serious tragedy.[2]

Just three months before, about sixty miles away, a sudden drop in water pressure caused highly toxic chemicals to back up into the drinking fountains in a New Jersey school, endangering hundreds of unsuspecting youngsters. Again, thanks to quick action and luck, no one was harmed.[2a]

After an outbreak of gastroenteritis in an Oklahoma school, public health inspectors found that none of the flushometer valve toilets with submerged inlets were provided with the vacuum breakers that prevent atmospheric pressure from forcing waste water into the supply lines. Each night, to conserve water and eliminate the possibility that classrooms might be flooded if a leak developed, the custodian turned off the valve from the main supply line. As the pressure in the supply lines was cut off, atmospheric pressure in the toilet bowls moved the waste water up to the drinking-supply reservoir. Most of the people affected were those who drank from the faucets on the first floor of the school; there were progressively fewer cases on the second and third floors as the atmospheric pressure moved less of the waste water to those heights.[8]

Though these are rather startling cases, they are not so unusual as you might think. The fact is that 95 million Americans drink water that is below federal standards or of unknown quality.[4]

A study of 969 water systems in the United States showed 41 percent were delivering waters of inferior quality to 2.5 million people, and 360,000 persons in the study population were drinking waters of potentially dangerous quality.[4a]

Even where average quality was good, the study, conducted by the Bureau of Water Hygiene of the Environmental Control Administration, found that occasional samples contained fecal bacteria, lead, copper, iron, manganese, and nitrates. Some of the very small communities were even drinking water on a day-to-day basis that exceeded the danger limits for one or more such chemicals as selenium, arsenic, and lead.

No book on food safety would be complete without a chapter on water. In the average community, over 140 gallons of water per person per day are used. Of this amount, about 60 gallons per person are used by each household, another 45 gallons by industry, 7 gallons by public services such as fire fighting, and 25 gallons by commercial firms. About 3 gallons are lost through leaks in underground water pipes.[5]

It takes 325 gallons of water to make one gallon of alcohol, 375 gallons to grow a one-pound sack of flour. The combined consumption of a cow, hog, sheep, and chicken will average about 40 gallons of water a day. And more than 77 billion gallons of water a day are pumped from rivers, reservoirs, and ponds in the widespread irrigation network that feeds western agriculture.[6] If contaminated water is used for any of these drinking and food-processing purposes, and proper treatment safeguards are not applied, the result may be human illness.

Another fact to emerge from the EPA study of water systems was that 79 percent of the systems were not inspected by state or county authorities in 1968, the last full calendar year prior to the study. In 50 percent of the cases, plant officials could not remember the last time state or local health departments had last surveyed the supply.[6a]

According to the study, 66 percent of the plant operators were inadequately trained in fundamental water microbiology, and 46 percent were deficient in chemistry relating to their plant operations.

How do you know whether the next glass of water you drink will be safe and pure?

The State Board of Health of Wisconsin, in a brochure directed at farmers, said: "People are likely to feel that odor-free, crystal clear, cold water, free from unpleasant tastes, is also safe water.

"True, these are all desirable qualities, but a water supply can meet these standards and still be as deadly as poison. Many a death has been caused by waterborne diseases. Many less serious diseases and intestinal disturbances are caused by impure water."[7]

In 1965, for instance, an outbreak of waterborne intestinal illness made 16,000 people sick in Riverside, California. Within a few days, several died, including an infant.[8] In the same year, an additional 24,000 cases of waterborne illnesses were reported in the nation. This, of course, is not a true representation of the total number of such illnesses, since, as in foodborne disease, there is faulty reporting.

Dr. Luther L. Terry, as Surgeon General of the Public Health Service in 1961, said: "When I entered the Public Health Service some twenty years ago, I was taught by the senior physicians of our Corps that whatever else in our environment was dangerous to health, our water supplies were safe. But three dangers—one potential and two actual—are causing us to re-examine our water's spotless reputation."[9] Dr. Terry listed virus diseases, new chemicals, and radiation.

Virus diseases are borne in human feces. Water inspectors determine the level of contamination of our drinking water by measuring *Escherichia coli,* which is present in fecal matter from humans and other mammals. Its presence is an indication of contamination.

Two families of viruses are excreted from the feces: the enteroviruses—which include polio, Coxsackie, and Echo viruses—and the adenoviruses. The enteroviruses multiply in the human intestines, and are discharged in the feces. Adenoviruses apparently multiply in the human intestines, too, and are also found in the feces, but less is known about them. Adenoviruses are associated with upper respiratory diseases causing inflammation of the mucous membrane of the respiratory system and the eyes.

The classic case of waterborne disease occurred in Chicago in 1933. As it was later proved, old, defective, and improperly designed plumbing fixtures and plumbing permitted the contamination of drinking water. As a result, 1,409 people came down with amebic dysentery. There were 98 deaths.[10]

This incident led to a tightening of water standards and expanded inspection, but public health officials today are still very much concerned about faulty plumbing. Cross-connections make possible the contamination of potable water, and are "ever-present dangers," the Public Health Service maintains.[11]

Why do such cross-connections exist?

One reason is, the Public Health Service says, that a connection is made by a plumber unaware of the danger. He does not realize that water flow may occur in a reverse direction or even uphill. Other reasons why such connections are made is the simple one of convenience, combined with a false reliance on a valve or other mechanical device or water pressure as adequate protection. Valves may fail or be carelessly left open.

Such cases are legion. One of the most famous occurred in 1932, when, during a five-week period, more than 10 percent of the 347 children in Huskerville, Nebraska, contracted polio. A study of the water supply revealed that the afflicted children lived in areas where flush-valve water closets lacked vacuum breakers.

But sewage in our drinking water is not the only danger. In the past twenty-five years, at least a half-million new chemical compounds have come into existence. Many of them are discharged into our rivers and streams or dumped on the ground to seep into our underground water supplies.

According to Dr. Luther Terry, the durability of some of today's new substances was demonstrated by the national water-quality headquarters in the Robert A. Taft Sanitary Engineering Center in Cincinnati, Ohio. One day a water sample from St. Louis produced a peculiar new mark on a routine infrared test procedure. A week later, the same mark appeared in a test made from water sent in from New Orleans. The chemical had traveled hundreds of miles in the Mississippi River without undergoing any change, and showed up in one liter of water taken from the billions of gallons of water that pass New Orleans every day.

Fish are good indicators of the dangers of polluted waters. Billions of fish have died in the past ten years from chemicals in our streams, rivers, lakes, and oceans.

Amazingly small amounts of pesticides can kill shrimp, crab, and other acquatic life. One part DDT in a billion parts of water was found to kill blue crabs in eight days. The relationship is the same as one ounce of chocolate syrup to 10 million gallons of milk.[12]

Nearly 25 percent of all DDT compounds produced to date may be polluting the oceans, according to a 1971 report by the Ocean Affairs Board of the National Academy of Sciences–National Research Council. The report predicts that more of these toxic compounds will eventually seep, drain or drift into the ocean, the "ultimate accumulation site" for persistent pesticides and other similar chemicals.

"Marine fish are almost universally contaminated with chlorinated hydrocarbon residues," the report states. For example, the ripe eggs of speckled sea trout on the south Texas coast contain about eight parts per million of DDT residue; five parts per million residue causes 100 percent mortality among young freshwater trout, and "the evidence is presumptive for similar reproductive failure in the sea trout," the report states. The number of speckled sea trout in the area has declined from 30 per acre in 1964 to 0.2 per acre in 1969.

Pesticides cause eggshell thinning and other harmful effects to birds, and "populations of fish-eating birds have experienced reproductive failure and decline." Toxic compounds can interfere with reproductive and growth processes of shrimp, crabs, fish, and other marine animals and cause declines in their populations. With continued accumulations of persistent pesticides in the marine ecosystem, the report warns, additional species of marine life will be threatened.

The warning is clear. What is bad for the fish is bad for man. Take the dumping of detergents into the water.

Water begins to foam if there is as much as one part of alkyl benzenesulfonate, a component of hard detergents, in a million parts of water. Hard ABS detergents were being dumped into the water supply by housewives and industry, and staying there. The Public Health Service allowed one half of one part of ABS in a million parts of water. When the foam choked rivers and streams, and people complained about turning on the faucet and having the water "foam" out, the federal government insisted that detergent manufacturers change to "soft" detergents, which do not foam.

However, according to the experts who prepared *A Strategy for a Livable Environment* for the Secretary of Health, Education, and Welfare in 1967, the new detergent compounds are suspected of killing large numbers of fish by affecting their eggs.

It has not been proved that detergents in water affect human health directly. However, in 1962, 1,727 incidents involving ingestion of detergents, as well as soaps and cleaners, in children under five years of age were reported to the National Clearinghouse for Poison Control Centers.

Dr. Jay M. Arena, of the Poison Control Center, Duke University Medical Center, said in the *Journal of the American Medical Association*, on October 5, 1964, in an article entitled "Poisonings and Other Health Hazards Associated with Use of Detergents":

"Because they are such necessary and familiar household items, cleaners are not usually regarded as hazardous substances. . . .

"Unfortunately, the terms 'soap' and 'detergent' are synonymous to many laymen, and I know of an incident, for example, in which significant injury occurred from the misuse of detergent instead of a soapsuds enema."

In explaining the toxicity of various detergents, Dr. Arena said:

"The cationic detergents or quaternary ammonium compounds are available commercially as ointments, powders, tinctures and aqueous solutions and are used to destroy bacteria on skin, surgical instruments, cooking utensils, sickroom supplies and diapers.

"The toxicity of quaternary ammonium germicides has not been definitely established, but the human fatal dose by ingestion has been estimated to be between 1 and 3 gm. . . . The principal manifestations of poisoning from ingestion of these agents are vomiting, collapse and coma. Specifically there is local and gastrointestinal irritation, restlessness, apprehension, confusion, dyspnea, cyanosis, convulsions, muscle weakness, and death due to paralysis of respiratory muscles.

"There are numerous anionic surface-active agents. All are only moderately toxic. . . .

"Packaged detergent granules: . . . manufacturers classify these products as (*a*) light-duty for dishes and baby clothes, (*b*) all-purpose, sudsing products for laundry and general use, and (*c*) washday, low-sudsing detergents made especially for automatic washers. However, with respect to their toxicity and the treatment of patients who have ingested them, all of these products may be considered as a single group. . . .

"Although the systemic toxicity of carbonates and silicates is low, their solutions can be very alkaline and may produce corrosive burning of the mucous membranes. . . . The ingestion of sodium tripolyphosphate and hexametaphosphate has caused severe gastroenteritis, with vomiting and diarrhea, and at least one child has developed esophageal stricture.

"Liquid detergents: . . . The formulations of the liquid household detergent preparations are similar to those of the detergent granules; the main difference is that they are in aqueous or hydroalcoholic solutions; . . . Manifestations associated with the swallowing of appreciable quantities should be no different than those expected from the ingestions of detergent granules. . . .

"However, several highly advertised general-purpose liquid cleaning agents contain petroleum distillates or pine oil made soluble by the addition of synthetic surfactants. . . . Aspiration of petroleum distillates, which may occur at the time of ingestion or later in association with vomiting, may produce hydrocarbon pneumonitis: involvement of the lungs has not been reported in cases in which the ingested product did not contain substantial amounts of this substance. Pine oil causes irritation of the eyes, mucous membranes, and gastrointestinal and genitourinary tracts, and depression of the central nervous system with hypothermia and respiratory failure. . . ."[13]

After a two-year Food and Drug investigation into dishwater detergents, the Food and Drug Administration began seizures of products its scientists said contained "hazardous chemicals." The detergents contain sodium metasilicate, a substance that is toxic when swallowed. It can cause severe eye irritation and has caused blindness in rabbits.

An ingredient in detergents—phosphate—has been identified as the factor that has choked our lakes and waterways with decayed algae, which has led to diminished oxygen and the death of the waters. The waterways have become uninhabitable for fish and unappealing for recreation.

Phosphate detergents have been banned in certain areas by local governments. If estimates are correct, about 280 million pounds of phosphorus per year would be withheld from surface waters if all detergents containing phosphates were banned. That would still leave at least 680 million pounds of phosphate that would enter our water to cause excessive growth of algae.[13a]

Ironically, it is very easy to rid sewage systems of phosphates before they reach waterways. They are easily precipitated by the addition of metal ions such as aluminum or iron in treatment plants.[13b]

In the rush to replace phosphates, eager detergent manufacturers substituted sodium nitrilotriacetate (NTA). It was discovered that NTA presented a danger far greater than the eutrophic effects of phosphates. On December 18, 1970, detergent manufacturers agreed at the request of the Surgeon General of the United States, to "voluntarily" suspend use of NTA pending further tests. The request was based on a report from the National Institute of Environmental Health Science that NTA may, in combination with a heavy metal, such as cadmium or mercury, cause birth defects. NTA increases the transmission of these metals through the placental barrier.[13c]

Here is a simple test to determine the presence of detergents or other organic materials present in a river, well, or glass of water. It was devised by George J. Crits, Assistant Technical Manager of the Cochrane Division of the Crane Company, Philadelphia.

Take a tall, cylindrical bottle similar to an olive bottle. Fill it half full of water, stopper it, and shake it. Presence of high amounts of detergents or soap will cause a noticeable foam. Small amounts will cause a film that will travel upward on the side of the glass bottle. The film rises until it disappears at a height dependent on the contamination in the water. The greater the contamination, the greater the height of the film or ring.

Perhaps the Public Health Service will reconsider its permissible limits for detergents, as it did for other chemicals in our water supply in 1962. Take arsenic, for instance. In the revision, the Public Health Service said:

"The widespread use of inorganic arsenic in insecticides and its presence in animal foods, tobacco and other sources make it necessary to set a limit on the concentration of arsenic in drinking water.

"The toxicity of arsenic is well known and the ingestion of as little as 100 mg. usually results in severe poisoning. Chronic poisoning from arsenic may be insidious and pernicious. A considerable proportion is retained at low intake levels. A single dose may require ten days for complete disappearance.

"Recent evidence supports the view that arsenic may be carcinogenic. The incidence of skin cancer has also been reported to be unusually high in areas of England where arsenic was present in drinking water at a level of 12 mg. per million parts.

"Formerly, the U.S. Public Health Service drinking water standards from 1946 established an arsenic limit of 0.05 mgs./1. In light of our present knowledge concerning the potential health hazard from the ingestion of inorganic arsenic, the concentration of arsenic in drinking water should not exceed 0.01 mg./1."

And take nitrate, which is deliberately added to food, including infants' food. The Public Health Service says:

"Serious and occasionally fatal poisonings in infants have occurred following ingestion of all waters shown to contain nitrate. This has occurred with sufficient frequency and widespread geographic distribution to compel recognition of this hazard by assigning a limit to the concentration of nitrate in drinking water.

"From 1945 to 1950, there were 278 cases of methemoglobinemia, including 39 deaths due to nitrate in drinking water." In the *Journal of the American Medical Association*, June 7, 1971, a detailed case was presented of an infant who suffered from methemoglobinemia because his milk formula was prepared with water containing excess nitrate.[13d]

"Breast-fed infants drinking from mothers who drank nitrate-contaminated water may be poisoned. Cows drinking water containing nitrate may produce milk sufficiently high in nitrate to result in infant poisoning. Both man and animals can be poisoned by nitrate if the concentration is sufficiently great."

Cadmium from zinc-galvanized iron and from electroplating plants also gets into our water. According to the Public Health Service, cadmium is a biologically nonessential, nonbeneficial element, and it is highly toxic:

"Recognition of the serious toxic potential of cadmium when taken by mouth is based on poisoning from cadmium-contaminated food and beverages, epidemiologic evidence that cadmium may be associated with renal arterial hypertension under certain conditions and long-term oral toxicity studies in animals," the Public Health Service said.

"Several instances have been reported of poisoning from eating substances contaminated with cadmium. A group of school children were made ill by eating popsicles containing 13 to 15 mg./1 part cadmium. All levels of dietary cadium so far tested have shown cadmium accumulation in the soft tissues down to and including 0.1 mg./1 in drinking water."

A United States Geological Survey reported on March 30, 1971, that seven toxic metals have been found in many of the nation's streams and lakes but that "dangerous concentrations were apparently rare." The survey found that of 720 samples taken in 50 states and the District of Columbia, 42 percent contained cadmium; 63 percent contained lead; 1.5 percent contained chromium; 37 percent contained cobalt, and almost 100 percent contained zinc.

And, just as scientists have found that individually harmless chemicals when combined in food or drugs may become toxic, they have also found the same thing in water. Water chemists refer to such combinations as "gunk."

Phosphate residues found in detergents, municipal sewage, and industrial wastes and agricultural products over-fertilize aquatic plant life and result in the process known as eutrophication. When water supplies are taken from lakes or streams that are undergoing eutrophication, the algae affect treatment plant operations, cause clogging of filters, may cause undesirable tastes and odor, and impair water quality.[14]

The old dangers from our water have subsided. The death rate from typhoid fever declined from 31.2 deaths per 100,000 population in 1900 to 0.01 in 1960.[15] But there are new dangers. The Public Health Service in its revised water standards of 1962, said: "The development of the nuclear industry has been attended by a small, unavoidable increase of radioactivity in the environment.

"The advisory committee, in considering limits which should be established for drinking water, recommended limits for only two of the nuclides—Radium 226 and Strontium 90.

"With Radium 226, above-average levels of intake generally occur only in unusual situations where the drinking water contains naturally occurring Radium 226 in greater-than-average amounts, as in the case of ground waters or from the pollution of the supply by industrial discharges of waste containing radium.

"The principal source of Strontium 90 in the environment to date has been due to fallout from weapons tests, and human intake of Strontium 90 to date has been primarily from food."

As the nuclear industry expands and the weapons race continues, the dangers from radioactive wastes in our water will increase. For instance, another radionuclide that was not included in the "United States Public Health Service Drinking Water Standards" of 1962, tritium, has been recognized as "of health significance." It is a radioactive isotope of hydrogen with an atomic weight of three, and a half-life of 1226 years. Tritium has the same chemical properties as hydrogen, and combines with oxygen to form water (^3H_2O). It cannot, by any standard treatment procedures, be separated from ordinary water. Extreme care is taken in the application of nuclear energy so that maximum permissible concentration of tritium in water will not be exceeded. This concentration at present is 0.003 microcuries per ml. above natural background.[16]

Modern man has also deliberately added chemicals to his water. Chlorine and fluoride are well-known additives, but water-softening chemicals are sometimes overlooked.

Congestive heart failure is a common cardiovascular problem today. Fluid retention in liver and kidney disease and in pregnancy is also widespread.

Dr. Peter Braun and Dr. Alvin A. Florin, of the New Jersey State Department of Health, writing on "Drinking

Water and Congestive Heart Failure" in the *Journal of the Medical Society of New Jersey*, in 1963, said: "The restriction of sodium in the diet has become a cornerstone of medical treatment in edematous [water-retention] states.

"Recent literature contains well-documented examples of sodium from drinking water as a cause of cardiac decompensation. In many areas, sodium in water supplies may be a significant factor in chronic congestive heart failure and other edematous states."

Drs. Braun and Florin said persons on a restricted salt diet should drink water containing no more than twenty milligrams per liter. But, they pointed out, commercial and domestic water-softening equipment can raise sodium concentrations to levels far beyond those desirable for such patients. In some New Jersey water supplies, the doctors found concentrations as high as 260 milligrams of sodium per liter.

In most American communities, public drinking water is one of the world's greatest bargains. The average cost of the sixty gallons or more of good water delivered into your home each day via the tap is a fraction of the cost of one bottle of soda.[17] But you may be getting more from your kitchen tap than you bargained for.

G. M. Hansler, Regional Assistant Administrator of Consumer Protection and Environmental Health Service, Health, Education, and Welfare, New York, emphasized the seriousness of the drinking-water problem in America today.[18] He said that in suburban communities in New Jersey, for instance, there were outbreaks of hepatitis and gastroenteritis after spring flooding in 1968. Drinking-water systems in these communities were suspect. Although unusually heavy rains can be construed as "acts of God," *proper* construction, maintenance, and routine surveillance of water systems can prevent such occurrences.

Mr. Hansler told of a new hospital in Mississippi that inadvertently had the water from its boiler control system connected with the drinking-water supply. People were drinking chromate ions in their water. He said this also happens as the result of air conditioners being cross-connected with water supplies.

At present, the level of chromate ions that can be tolerated by man for a lifetime without adverse effects on health is unknown. It is known that, when inhaled, chromium causes cancer.[19]

No section of the nation is without at least one major water problem, either in distribution, supply, quality, pollution, floods, or variability.[20] And the situation is going to get worse. By 1980, we are expected to use almost 600 billion gallons daily of fresh water, but only 515 will be available.

As the levels of water in our streams, rivers, and reservoirs goes down, pollution goes up. Less open land for natural purification and more chemicals from our farms and industries will compound the problem. More people will make more sewage. We shall have to begin using the same water over and over again. Yet, even now, viruses can escape our present water-purification and disinfection processes.[21]

But with all the talk about water pollution, relatively little research is going on concerning the health effects of the metals, chemicals, sewages, and viruses in our drinking water.

7 NOTES

1. G. M. Hansler, Acting Assistant Regional Director, Environmental Control Administration, U.S. Health, Education, and Welfare Department, New York region. Tape-recorded interview with author, July, 1968.
2. Ruth Winter, "Even Cyanide Gets Into the Water," *Newark Star-Ledger,* June 18, 1969.
2a. *Ibid.*
3. *Ibid.*
4. *A Strategy for a Livable Environment,* a report to the Secretary of Health, Education, and Welfare by the Task Force on Environmental Health and Related Problems, June, 1967.
4a. Survey, 1969, conducted by the Bureau of Water Hygiene, Department of Health, Education and Welfare, Environmental Health Service.
5. U.S. Department of Health, Education, and Welfare, "Water Supply Activities," Publication No. 1511, 1966.
6. Howard Earl, "Water Pollution, That Dirty Mess," *Today's Health,* March, 1966.
6a. *Ibid.*

7. State Board of Health Publication, Wisconsin.
8. U.S. Department of Health, Education, and Welfare, "Water Supply Activities," No. 1511, 1966.
9. Luther Terry, M.D., Surgeon General of the U.S. Public Health Service, "Let's Stop Poisoning Our Water," in *This Week*, December 3, 1961.
10. "Water Supply and Plumbing . . . ," HEW Publication No. 957, 1963.
11. *Ibid.*
12. Howard Earl, *op. cit.*
13. Jay M. Arena, M.D., Poison Control Center, Duke University Medical Center, "Poisonings and Other Health Hazards Associated with Use of Detergents," *Journal of the American Medical Association*, October 5, 1964, Vol. 190, pp. 56–58.
13a. "Cleaning Our Environment: The Chemical Basis for Action," a report by the Subcommittee on Environmental Improvement, American Chemical Society, Washington, D.C. 1969, p. 151.
13b. Allen Hammond, "Phosphate Replacements: Problems with the Washday Miracle," *Science*, April 23, 1971, pp. 361–363.
13c. *Ibid.*
13d. Louis W. Miller, M.D., "Methemoglobinemia Associated with Well Water," *Journal of the American Medical Association*, June 7, 1971, Vol. 216, No. 10, p. 1642.
14. Dr. Gerard Rohlich, Professor of Civil Engineering and Director of Water Resources Center, University of Wisconsin, on "Eutrophication," presented to 41st annual convention, Soap and Detergent Association, New York, January 25, 1968.
15. "Patterns of Disease," Parke-Davis, July, 1963.
16. Atomic Energy Commission, *Regulations,* 10 CFR, Part 20.
17. U.S. Department of Health, Education, and Welfare, "Water Supply Activities," No. 1511, 1966.
18. G. M. Hansler, interview cited above.
19. "Public Health Service Drinking Water Standards Revision, 1962.
20. *Patterns of Disease*, July, 1963.
21. Luther Terry, *op. cit.*

8

Eating Out Tonight?

As I stood in line at a barbecue in an exclusive tennis and swim club whose members are among the most intelligent and well-to-do in New Jersey, the French chef behind the table, cigarette dangling from his mouth, was dishing out spaghetti with his fingers. He was sweating profusely. An elderly gentleman in front of me complained loudly at such an obvious breach of sanitation, and stalked out of the club. I stayed, fascinated by a flagrant violation of the most elementary health practices involved in serving food to the public.

Farther down the line, I saw a young assistant spooning desserts on a plate, and stopping every once in a while to lick his fingers. The waitresses, hired for the summer, had long hairdos that hung loose over the food they served. Flies were plentiful. I looked at the silverware; and some of it was encrusted with food. A meat sandwich that was served had obviously been sliced from meat that had been cooked the day before, for it was dry and unappetizing.

I was reminded of what I had learned from food inspectors: much of what I saw actually goes on in many of the more than half a million eating places in the United States. The difficulty is that the patrons just don't see it, for most of it takes place behind the scenes, in the kitchens and larders.

We Americans buy 28 billion meals a year, and more than 540,000 eating and drinking places serve approximately 78 million customers daily. That number of patrons represents nearly half the population in the nation.[1]

The Bank of America did a two-year survey, and found

that the business of serving food totals $27 billion at retail annually. Separate eating and drinking places now rank as the Number One type of retail outlet in the nation. "They outnumber grocery stores. They outnumber service stations," the survey said. "American families now spend $400 to $500 annually. In fact, we are now spending more to 'eat out' than we do to buy new and used automobiles."

Yet the survey quoted recent figures, compiled by Dun & Bradstreet, indicating that over half of the restaurants in the United States show no taxable profit. Each year, the survey said, eating and drinking places account for around 20 percent of all retail failure. Only half of all food-service operations maintain the same ownership for five years or more.[2]

But such financial perils may not be the only hazard. The Public Health Service, in a publication directed at food-service personnel, said:

"Yours is a dangerous business, . . . dangerous because food and drink that are not carefully handled and prepared too often are a source of disease—and death.

"A single germ in 24 hours will under favorable conditions produce 281,000,000,000,000 other germs capable of doing the same.

"People have died because they ate and drank heartily, trusting in those who prepared and served the food. And people will continue to sicken and die because they eat out, unless you protect them. You cannot always tell by tasting whether food is safe or not."[3]

The food service industry is the fourth largest in the United States, involving over a half-million establishments and employing over three million people. Changing patterns in social and economic life have led to more and more meals eaten away from home.[4]

Thirty-two percent of foodborne illnesses reported began in the kitchens of commercial eating places.[4a]

It was brought out at an American Public Health Association-sponsored conference on food protection, in April, 1971, that an extremely high percentage of the managers and employees in food-service establishments have not had any formal training in the principles of microbiology, and therefore lack the basic knowledge needed to recognize microbiologically hazardous conditions when they occur in their individual work situations. Extensive efforts on the part of regulatory agencies during and immediately following World War II to provide on-the-job training in the basic principles of food protection failed to achieve any significant change in handling practices, and were ulti- mately discarded.

According to the report of the conference's panel on "Prevention of Mishandling of Foods in Commercial and Institutional Food Service Operations": Over the years, many regulatory agencies have espoused the philosophy that "education" is a more effective tool in securing com- pliance with food service sanitation laws than is strict enforcement. One need only observe the flagrant violation of many food protection requirements, such as the keeping of potentially hazardous food at unsafe temperatures, to be convinced that this approach has not worked as effec- tively as it was hoped it would.

"The shrinking tax dollar together with mounting public pressures for prevention of environmental pollution have placed food protection programs in fiercer competition with other environmental programs such as air, water, and solid- waste programs. A lack of imaginative planning and de- velopment of improved program techniques have mini- mized program accomplishments. General public apathy to insanitation and mishandling practices in food service estab- lishments have further complicated the problem of securing support for food sanitation programs at all levels of govern- ment.

The Public Health Service does have thick files on outbreaks of food poisoning, although personnel admit that their records are just the very tip of the iceberg. Such illnesses are not all from germs. Take the case of the barbecue-stand patron who loved chili. He ate some "mexihot." Six hours later, he was dead. On the same day, when several other patrons became ill, food sanitarians traced the case to a roadside stand. The chef refused to believe it, and ate some chili himself to prove his point. He died.[5]

Though there were no disease germs in the food, the laboratory discovered a quantity of sodium fluoride, a poison used for killing roaches, in the chili. In another incident, 236 people on the West Coast got sick because roach powder had been mistaken for powdered milk and mixed with some scrambled eggs. Forty-seven died.

Though most of the restaurant slip-ups are not so dead- ly, they are far more prevalent. The East Orange, New

Jersey, Health Department has a budget for four inspectors (which means all the positions are not always filled). The inspectors also investigate dog bites, ragweed pollution, swimming pools, sewage, garbage, pet shops, gasoline stations, citizens' complaints, rodents, smoke, and other nuisances. The population of the city is 77,259.

During one year, 622 inspections of 305 East Orange food establishments were made; an average of five violations of the Sanitary Code was found on each inspection.

What exactly does an inspector look for when he goes into a restaurant?

According to William Clinger, Regional Sanitation Consultant, Environmental Health, United States Public Health Service:

"The most frequent violation concerns temperature. The cold foods are not kept under proper refrigeration and the hot foods are not kept hot enough. Refrigeration should be below 45 degrees. Steam tables should be 140 degrees.

"The second most common violation is improper storage. Flour and onions stored on the floor, for instance.

"Inspectors look to see whether foods which are displayed are protected by glass ... whether cream pies and custards are refrigerated.

"Does food splash on single service items such as paper cups? This is a problem today.

"Are there open sugar bowls? Sugar should be served in individual packages or in a closed container.

"What are the toilet facilities like? Can help wash their hands in a convenient place?

"There should be no cross-connections with or back-siphonage into the water supply from sinks, dishwashing machines, lavatories, or toilets.

"A frequent violation is that the drains from the steam tables reflush over the tables.

"Grease traps should be cleaned to avoid clogging and should be in good repair.

"Vapona insect strips are used improperly in many places today. They are sometimes hung over tables used for preparation of food, and condensation from the heat causes them to drip into the food and milk.

"No food should be left overnight.

"Garbage should be covered and disposed of promptly.

"Insect and rodent killers should be kept far away from the food.

"Dishes should be washed either by hot water in a dishwashing machine or disinfected with chemicals.

"The inspector must know the source of the restaurant's food supplies.

"There should be no smoking while food is being prepared.

"Employees should wear headbands, hairnets, or caps, and all uniforms should be clean."[7]

Restaurant inspection itself is a controversial thing. When Dr. George Kupchik, Director of Environmental Health of the American Public Health Association, was asked by the author whether he felt confident when he went into a restaurant that was not inspected by a health department, he said, "I don't feel confident when I go into a restaurant that I know is inspected! Restaurants inspected every three months for ten years are no better than restaurants that have never been inspected," he said. "There . no question in my mind that some of the fanciest-name restaurants lack adequate sanitation. On the other hand, some small, family-run eating places have excellent sanitation."[8]

The fallibility of inspection was demonstrated in an item from *FDA Papers* in February, 1967:

"Two Philadelphia FDA inspectors were too close to a recent food poisoning case for comfort. Wayne Stafford and Kenneth Martin developed severe symptoms of food poisoning after eating at a diner in Allentown, Pa., in October. Inspector Stafford required hospital treatment. Investigation showed that the diner used drinking water from a contaminated well and stored and handled food in an insanitary manner. Bacteriological findings in sausage and butter samples taken at the diner were reported to the State of Pennsylvania Department of Health."

The question of health examinations for food handlers is highly controversial.

"In 1921," Dr. Aaron Haskin, Health Officer for Newark, New Jersey, explained, "the City of Newark passed an ordinance requiring the physical examination of food handlers, the theory being that a sick food handler would transmit disease through food.

"While the theory is commendable, the actual practice cannot be considered so. The physical examination con-

sisted of an X ray of the chest, blood serology, and a
mouth smear. The examination also included a very curso-
ry screening for superficial skin infections. If we realize
that the diseases most apt to spread through food are of
the salmonella, typhoid and dysentery groups and the
staphylococcus and streptococcus organisms, you can readily see that the type of examination as carried out is not
directed at these organisms; the program has degenerated
into a case-finding program for tuberculosis and venereal
disease.

"The fallacy of the physical examination for the prevention of the spread of disease through food becomes more
obvious when we realize that even if we were able to
carry out the proper type of examination, it would only
show that at the time of examination the food handler was
well. But the next day he could easily become infected
with any of these important diseases, and be a source of
infection, and still have in his possession a health card
from a health department testifying that he is well and
that it is safe for him to handle food.

"In my opinion, this gives the public and the food
handler a false sense of security, because unless that food
handler is acquainted with the disease that he may have
and unless he knows how to properly conduct himself so
that he may not infect the food, the public is not safe. I
believe that no physical examination of the food handler is
the answer to this problem and certainly the public should
not be lulled into this false sense of security.

"In 1951, I made an evaluation of the program of food
handler examination and I found that we spent $40,000 to
$50,000. As a result of the expenditure of this amount of
money, we found two cases of tuberculosis not heretofore
known and several cases of syphilis.

"In view of the fact that this had turned out to be a
case-finding program in tuberculosis and syphilis, I was
sure that the $40,000 to $50,000 spent in a different
manner for case finding would discover a great many
more cases."

Dr. Haskin said that when he abolished food-handler
examinations in 1952, he was warned that dire consequences would fall upon the citizens of Newark. Instead
of the examination, however, Dr. Haskin has required all
food handlers to go through a week's course in sanitation.
The course informs the food handler of the types of

diseases he may have that will spread through food, chiefly the diarrheal diseases, staph and strep infections and Vincent's infection of the mouth. The food handler is taught how these diseases appear on the body, how he as a food handler may infect foods with these various organisms, and how he as a food handler could prevent the infection of food. The food handlers are taught good sanitation practices, including dishwashing and hand washing.

Dr. Haskin also pointed out that a survey conducted by the Sanitation Foundation found that the average intelligence of a food handler is that of an eighth grader. He said food handlers' training programs have to be adapted to their IQ level.

Dr. Haskin believes that the Newark program is working well and that no food handler should be permitted to operate unless he can show a certificate testifying that he has successfully passed this form of instruction. He said his inspectors ask to see the certificates when they go into a food establishment.[9]

East Orange Health Officer Robert Lackey said that food handlers in his jurisdiction are required to have X rays of the chest only. He expressed doubt that this was of much use. He did point out that many food handlers in American restaurants today are immigrants from other countries, from Puerto Rico, and from the American South. He said that many of them have parasites that can and are being transmitted to the food they handle. He said experience has shown that even if you do rid the food handlers of their worms and other parasites, they soon become reinfected by members of their own households.

Mr. Lackey thought that perhaps stool cultures might be of some help in detecting food handlers carrying parasites.

Dr. Haskin, on the other hand, said, "If we did stool cultures on all food handlers in Newark, our laboratory would be a cesspool, and the people still wouldn't have adequate protection."

Mr. Lackey believes that education of food handlers is very difficult to implement. He said that the turnover of restaurant help is almost 100 percent. "If a town has one thousand positions for food handlers, two thousand will have filled them in a year," he explained. He feels that vocational schools will have to play a larger part in the education of persons who intend to work in food-handling jobs.[10]

In the *Journal of the American Medical Association*, on November 6, 1967, in answer to a Florida physician's question about food-handling certificates providing a "false security and totally inadequate control of communicable disease," Dr. Robert W. Harkins of Chicago answered for the AMA:

"Public Health officials can do much to encourage food sanitation by providing educational services for both the manager and the employee. Periodic, realistic inspection of all food service operations can pinpoint weak gaps. A sense of urgency among public health officials must be conveyed to the owners, and it should be conveyed to the courts as they levy penalties on those establishments which fail to maintain adequate standards. Everyone wins when food service sanitation is high—the customers, the employees and the owner."[11]

A New York City Health Department inspection of 845 Manhattan restaurants—some of them world famous—showed that 86 percent were violating the city's Health Code.[11a]

As a result, the department will disclose the names of restaurants that do not promptly correct violations, and if they continue thereafter, the restaurant will be closed.

Also, the Health Department is developing a posting system for New York City similar to the one used in Atlanta, Georgia. In that city, for the past twenty years, restaurateurs must keep scorecards of their latest inspection, showing how they rated in food supplies, food protection, personnel, equipment and utensils, water supply, waste disposal, toilets and lavatories, vermin, and plumbing. Each category is rated on a scale of 1 to 5. The scorecard must be shown to patrons on demand.

I asked food inspectors what they would not order when they go into an unfamiliar restaurant. Most agreed that they would not order raw shellfish unless they knew that the product came from uncontaminated water. They would not order egg salad anywhere, and would be hesitant about ordering any cooked salad such as potato or chicken salad. Many were even cautious about ordering ham sandwiches because they said cooked ham is a frequent source of foodborne infection.

A prominent cookbook author told me that she would never order chicken salad or lobster thermidor unless she knew the chef and he liked her. She said leftover chicken may be used for the salad, and the shells of lobsters may be taken from the plates of patrons who had whole lobsters, and kept around the kitchen for several days before being filled.

Based on the opinions of the experts, here are the things you should look for when patronizing a restaurant:

● Is the outside clean? Are the bathrooms clean? Although different personnel may handle food, these factors are a good indication of whether the management insists on good sanitation.

● Are the dishes chipped? All equipment—including utensils and dishes—that touches food or drink should have a hard, smooth finish.

● Is garbage exposed? Are the counters and tables dirty?

● Are there flies or evidence of insect and rodent contamination?

● Are the food handlers smoking? Ashes from cigarettes or cigars and contamination from saliva are, of course, unsanitary.

● Is the restaurant in the basement? If it is, there is a greater possibility of contamination by flooding, sewage, and rats.

● Are the uniforms of the food handlers and servers clean? Do they wear caps to keep their hair from falling in the food?

● How do the waiters and waitresses set the silverware down? How do they serve the food? Are their fingers over and in everything?

● Are the glasses and silverware clean? Is there lipstick on the glasses or are there coffee grounds in the cup? (Water spots are not an indication of contamination.)

● Does the cream for the coffee feel warm? If it does, it is a good indication that it has been kept at room temperature and that temperature and time control are not observed by the management.

● Is the help coughing, spitting, or sneezing near the food?

● Are custards and cream-filled foods refrigerated?

● Are foods kept covered until they are served?

● Are the floors clean?

● In a restaurant other than a very small one, does a busboy remove your dirty dishes? Contamination from dirty plates may contaminate your food by way of the hands of the server if busboys or proper sanitation are absent.

● When your food is served, does the server use a utensil instead of his hands?

● Are the menus dirty?

● Do the waiters or waitresses have sores on their skins or do they have improperly bandaged fingers if they have been injured?

● Are the salads warm or wilted?

THE CATERED AFFAIR

One day in June, 1965, health officials in the Washington, D.C., area became aware of an outbreak of diarrheal disease. Some of Washington's most important citizens were affected. Investigation showed that those afflicted had attended one of nine catered parties held on June 5 or June 6. At each of the catered affairs, a variety of foods from the same caterer had been served. Of approximately 570 persons who had attended the parties, 431 were interviewed. One hundred and ninety-eight—45.9 percent—reported an illness that began at a mean interval of twenty hours after the festivities.

The most common symptoms included moderately severe diarrhea, abdominal cramps, and low-grade fever. Ages of the victims ranged from three to eighty years. One patient was hospitalized, and sixty-five required the services of a physician.

Isolated both from the foods served and from sick persons who had attended the party were *S. meleagridis*, *S. chester*, and *S. tennessee* forms of *Salmonella*.

In an effort to determine the extent of the secondary spread within the households of the affected, a telephone survey was made. Only one possible secondary case showed up. The husband had attended the party and had had severe diarrhea for several days. The wife had not attended the affair, but one week after her husband became ill, she too had the same symptoms.

An additional 142 illnesses not related to any of the parties were reported following newspaper publicity concerning the outbreak.

It was learned that prior to the June outbreak, several cases of salmonellosis related to the delicatessen in question were known by health authorities. In fact, stool specimens obtained from employees at the time were taken. Seven of the food handlers behind the serving counter were found to be excreting *S. meleagridis*. All denied any clinical illness. The employees were removed from their jobs and were treated with antibiotics. They were instructed not to return to work until three successive rectal cultures were negative. On June 2 and 3, just before the outbreak, four of these employees known to have been actively excreting the organism were working nevertheless. In all, a total of 356 known cases of salmonellosis occurred between May 10 and June 7, 1965.

The enterprise involved was a combination caterer-delicatessen-restaurant. It was physically divided into a delicatessen store, dining area, kitchen, bakery, second-floor banquet room, and basement storage area. It catered up to ten parties a month, and handled a large-volume restaurant delicatessen trade. It was estimated that more than four thousand people consumed food there between June 4 and June 7.

Following the closing of the firm by health authorities on June 7, of the 103 specimens cultured from employees, fifty-seven were found to be positive for *S. meleagridis*. Employees were treated with antibiotics. In the meantime, a complete environmental investigation of the premises was made. Here are some of the blatant violations found:

1. Lack of hand-washing facilities.
2. Inadequate employee rest rooms.
3. Presence of many cockroaches and flies.
4. A malfunctioning automatic dishwasher.
5. A nonrefrigerated delivery truck.
6. Inadequate refrigeration facilities.
7. Inadequate routine cleanup of food-processing areas.
8. Corned beef was cooled in a kitchen sink that had its drain plug situated several inches below a stationary grid.

9. Raw foods were not kept separate from cooked foods.

10. A can of frozen eggs was found in a temperature of 80 degrees.

11. Display cases in the delicatessen were used for cooling rather than holding, which placed a burden on the refrigeration units.

12. The slicers were contaminated, resulting in cross-contamination of sliced meat such as salami and bologna.

This case was reported in the *American Journal of Public Health*, in April, 1968, by Arnold Kaufmann, D.V.M.; Charles R. Hayman, M.D., M.P.M.; Frederick Heath, M.D., M.P.H.; and Murray Grant, M.D., D.P.H.[12]

The catering business, health officials admit, is one of the weakest links in food-protection efforts in the United States. The majority of local and interstate caterers are not inspected, and can operate without supervision.

Vernon E. Cordell, Director of Public Health and Safety for the National Restaurant Association, said that the proliferation of multiple-unit operations, franchised operations, and food services performed by management contractors had brought the food industry up against the problem of lack of uniform sanitation regulations.

Instances of poisoning from all sorts of catered affairs are numerous, and involve large numbers of people. For instance, the New York City Health Department reported that three thousand persons were made ill between April 5 and 11 in 1967. They attended catered affairs in the area. The desserts for these parties were made by Country Club Frozen Desserts, New York. The firm immediately shut down and ceased production and distribution until proper equipment and processing modifications could be made. The contamination was ultimately traced to nonpasteurized, sugared egg yolks used in the desserts, and manufactured by the Manhattan Egg Company, New York. The firm halted distribution of nonpasteurized egg products. The eggs were linked to other food-poisoning outbreaks in New Jersey, and were suspected of causing similar ailments in Massachusetts, Connecticut, and Michigan.[13]

Private clubs, fraternal orders, societies of various types, churches, and religious-affiliated organizations are not closely inspected. In fact, the church supper has become one of the most common events for such unwelcome guests as salmonella and staph poisons.

Current practices relating to this category of food services varies throughout the country. Some health agencies include such organizations in their inspections. Others exempt private clubs, church-affiliated, or certain educational organizations from permit or license requirements and regular sanitary inspections. "When this is the case," the Public Health Service says, "a rather substantial segment of the population receives no official health protection in this area of food control."[14]

East Orange Health Officer Robert Lackey also points out the difficulty of inspecting delicatessens and caterers of any sort. "The health agencies do not have the manpower or the money to bring these under surveillance," he said. He gave as an example cooked salads:

"Potato salad is brought in 50-pound containers from Brooklyn or Pennsylvania; and, historically, when we tested it we found contamination. Even though these products travel interstate, they are not necessarily inspected, and we do not have the inclination to go to another state and inspect the processing plant."[15]

In May, 1965, New Jersey started a sampling and inspection study of wholesale manufacturers and distributors of potentially hazardous foods, with particular attention paid to potato salad, macaroni salad, and coleslaw products. According to Robert E. Johnson, Bureau of Food and Drugs, New Jersey:

"Human contamination has been evident in the preparation and subsequent handling of these foods. We have related causes for high laboratory results of analyses to insanitary processing practices, use of contaminated ingredients, too slow cooling after preparation, improper storage facilities, improper repacking procedures and/or improper transportation facilities."

He said that in one study New Jersey did of ninety-four samples of potato salad, 35 percent had bacteria counts of 100,000.[16]

Some food inspectors suggest that a person who wants to give a private party, or who is in charge of arrangements for an organization, first get in touch with the local board of health to determine if the caterer has a license. Big caterers usually have good equipment, including temperature-control trucks, experienced help, and good sanitation know-how. But it pays to ask questions. In the meantime, after checking with the local health department, readers should see that potentially hazardous food is not exposed to improper temperatures during the party and that food is not contaminated by improper handling, either by the food handlers or by the guests.

VENDING MACHINES

One afternoon, a woman having her nails done and the manicurist doing her nails both did not feel well. Neither wanted to say anything, but as they sat opposite each other they both began to feel nauseated. Finally, the customer, with nails half polished and still wet, stood up from the table and rushed to the ladies' room. She was soon followed by the manicurist.

The fact that both had become ill almost simultaneously was not a coincidence. A chain of well-known restaurants had placed a shop containing automatic vending machines on the Short Hills, New Jersey, mall. The mall contains branches of many Fifth Avenue, New York, stores, and caters to a clientele drawn from many of the lush suburbs in the state. Both the customer and the manicurist had eaten sandwiches at noon from the vending machines. As with such illness, although typically foodborne, no report was made of the possible source of the illness to the local board of health. Perhaps there were too many such illnesses or the operation proved unprofitable, but the shop was closed and a conventional restaurant established in its place.

Though sales of products from vending machines are now in the billions, vending machines, like catering establishments, are a very weak link in the food-inspection protection program. Actually, vending machines are not new. Their existence has been recorded since a machine dispensed holy water in Alexandria, Egypt, in 200 B.C.[17] Vending machines first made their appearance in America in 1880. During the Second World War, as the shortage of all food-service personnel became acute, vending machines multiplied to fill the gap. There were no regulations concerning the vending of food, and it wasn't long before public health officers found that the machines were a source of trouble. Fruit juices, syrups, and other products would leave residues on the machines and in the machines and contaminate the fresh products. Delivery tubes were not sanitized, and microorganisms flourished in such an environment.

In the early fifties, a number of persons were made violently ill after drinking soda from vending machines. Defective valves had permitted carbon dioxide to back up into the copper water valve. Copper was leached, thus contaminating the soda, and the unwary drinkers suffered copper poisoning.

The British have been more aware of, and stricter with, potential hazards from vending machines.[18] In 1966, the *Public Health Inspector*, a British publication, revealed that there had been a great deal of trouble and many complaints concerning unsound meat pies and meat sandwiches purchased from vending machines. Upon investigation, they found that the delivery truck had no form of temperature or humidity control. Though pies and pastries were put into preheated, heat-controlled containers, no thermometers were used. The vending machines were cleaned in a perfunctory way—when time and staff were available. Internal lighting in the machines raised the temperature more than 40 degrees.[19]

In *The Medical Officer,* another British publication, in September, 1961, health officers recognized a potential hazard we have yet to see. They wrote:

"With the increasing habit of chewing gum, slot machines for dispensing gum have multiplied. Sometimes the operators would drive up in the night, remove the top, collect the money, and put new gum in it. During the later months of 1960 and 1961, a series of machines were brought before the justice of the peace.

"The slot machines dispensing unwrapped ball gum were examined bacteriologically. Three out of 60 were found to be contaminated by human bacteria. *Streptococcus pyrogens* lived for a week or 16 days. Staph and *E. coli* survived 33 and 40 days.

"Any machine with unwrapped sweets into which children can poke their fingers is fundamentally unsound."[20]

The English no longer permit the dispensing of unwrapped sweets in vending machines. The Americans do.

How many American children have picked up "one of those viruses" from a vending machine?

An Evaluation of Public Health Hazards from Microbiological Contamination of Foods, a Report of the Food Protection Committee of the Food and Nutrition Board, National Academy of Sciences, said of vending machines:

"The industry has, by self-regulation and education and by use of public health consultants and outside evaluation agencies, established a good operating control of food-vending machines.

"Recent findings cause some doubt on the efficiency of the lower temperature in restraining the growth of potentially hazardous bacteria. The upper level is, of course, insufficient to prevent growth of some thermophilic bacteria which, while not dangerous per se to health, could render the food unattractive. However, the primary risk in the case of food-vending machines seems to be related to improper processing and mishandling of foods prior to introduction into the machine, or to malfunction of the machine itself so that temperatures rise above 10 C. or drop below 66 C., even though machines are provided with controls designed to render them inoperative automatically when this happens."[21]

Public Health Service standards state:

"Vending machines dispensing potentially hazardous food shall be provided with adequate refrigerating or heating units, or both, and thermostatic controls which insure the maintenance of applicable temperatures at all times. Such vending machines shall also have controls which prevent the machine from vending potentially hazardous food until serviced by the operator in the event of power failure or other conditions which result in noncompliance with temperature requirements in the food-storage compartments.

"The machine location shall be such as to minimize the potential for contamination of the food, shall be well lighted, easily cleanable and shall be kept clean. Conveniently located handwashing facilities shall be available for use by employees serving or loading bulk food machines."[22]

Have you seen machines that do not follow this description?

What can the customers do to protect themselves and

their families against contaminated vending machine products?

"Nothing," said Dr. George Kupchik. "It is a closed mechanism. You simply have to rely on the operator."

But why not post a sign stating when the machine was last inspected, just as in elevators? Why not a list of standards for the product sold by the machine, and a visible temperature gauge? If the machine is supposed to dispense hot food above a certain temperature, why not say so, and have a thermometer that can be read by the buyer?

IT CAN BE DONE

Harold Simpson, a former chef, is now Corporate Director of Purchasing Control for Restaurant Associates Industries, an organization that owns everything from vending machines and snack bars to some of the best gourmet restaurants in the world, such as The Four Seasons and Forum of the Twelve Caesars in New York. He said:

"We have self-inspection and we hire two professional firms to check our sanitation, in addition to the checks by local and federal inspectors.

"We run our own classes for our food handlers, and the City of New York has a food-handling course in sanitation. There is no substitution for proper food-handling techniques.

"We also reject a lot of food if it does not meet our standards, which are very rigid," he continued. Mr. Simpson says that their standards are so rigid that they have not found a frozen-food manufacturer that measures up to their requirements.

"Some restaurants have gone into portion-cut, prepared convenience foods. This is a costly thing in itself, but they think it will reduce their own labor costs. We haven't been able to find the quality and the portion cuts we want. Furthermore, if all restaurants buy food from the same frozen-food manufacturer, then all restaurants would be alike."

Mr. Simpson, who was once a chef at the Greenbrier, Mount Washington, and Boca Raton hotels, said the reasons restaurants go under financially is poor location and lack of business knowledge:

"A man can be a hell of a cook, but without the other knowledge he won't succeed. People have to be trained in temperature control, how to care for perishable foods, how to rotate food items, how to prepare a menu.

"With the number of meals being served in this country, the number of bad results is very small," he said.[23]

No one wants to poison a patron, but even the most efficient operation can be hazardous. An airline wanted to serve gourmet meals on its flights. Meals prepared by approved processors were frozen; there were temperature controls on each "stack" of meals, and the trucks that carried them were properly temperature-controlled, as were the kitchens aboard the planes. However, no one could predict which meals the passengers would order. Some meals were taken on and off airliners thirty-two times, often with a loss of temperature control. A potentially hazardous situation developed, and the airline lost more than a million dollars, finally changing back to the old system.[24]

Several federal food inspectors and I were discussing the hazards of eating in some restaurants while traveling by car. They pointed out that the FDA actually has legal jurisdiction over places along the highway because they can inspect food that is used in interstate commerce. They wondered whether it would not be possible for federal inspectors to give a "seal of approval" to tourist places that qualified, saying they were sanitary.

The value of such a seal is obvious, and if you think so too, and have had some bad experiences while traveling, you could encourage such a step by writing to your congressman.

8 NOTES

1. "From Hand to Mouth," U.S. Department of Health, Education, and Welfare, Public Health Service Publication No. 281.

2. Report of Bank of America Survey, *Newark News*, July 7, 1968.

3. "From Hand to Mouth," *loc. cit.*

4. Preliminary Position Papers, National Conference on Food Protection, Denver, Colorado, April 4–8, 1971, sponsored by the American Public Health Association.

4a. Center for Disease Control, Atlanta, Georgia, "Foodborne Outbreaks," Annual Summary, 1970.

5. "From Hand to Mouth," *loc. cit.*

6. Board of Health, City of East Orange, New Jersey, Annual Report, 1966.

7. William Clinger, tape-recorded interview with author, July, 1968.

8. Dr. George Kupchik, Director of Environmental Health, American Public Health Association, tape-recorded interview with author, February 26, 1968.

9. Aaron H. Haskin, M.D., Health Officer, Newark, N.J., "Physical Examination of Food Handlers Is Not a Useful Requirement," *Public Health News*, New Jersey State Department of Health, June, 1967. Also, author's interview with Mr. Haskin in July, 1968.

10. Robert Lackey, East Orange Health Officer, tape-recorded interview with author, June, 1968.

11. Robert Harkins, Ph.D., Chicago, "Food Handling Certificates," *Journal of the American Medical Association*, November 6, 1967, Vol. 202, No. 6.

11a. John Sibley, "Health Grading of Restaurants Urged," *New York Times*, June 9, 1971, p. 45.

12. Arnold Kaufman, D.V.M.; Charles Hayman, M.D.; Frederick Heath, M.D.; Murray Grant, M.D., "Salmonellosis Epidemic Related to Caterer-Delicatessen Restaurant," *American Journal of Public Health*, April, 1968, Vol. 58, No. 4.

13. *FDA Papers*, July–August, 1967, p. 35.

14. *Food Service Sanitation Manual*, 1962, Public Health Service Publication 934.

15. Tape-recorded interview with author, June, 1968.

16. Robert E. Johnson, Principal Sanitarian, Bureau of Food

and Drugs, New Jersey Department of Health, "New Jersey's Program to Control Quality in Potentially Hazardous Foods," *Public Health News,* June, 1968, Vol. 49, No. 6.

17. *Public Health Reports,* October, 1958.
18. *Ibid.,* April, 1957.
19. *Public Health Inspector,* November, 1966.
20. *The Medical Officer,* September, 1961.
21. *An Evaluation of Public Health Hazards from Microbiological Contamination of Foods,* a report of the Food Protection Committee, National Academy of Sciences and National Research Council, Publication 1195, 1964.
22. *The Vending of Food and Beverages,* 1965 recommendations of the Public Health Service.
23. Harold Simpson, Corporate Director, Restaurant Associates Industries, interview with author, July, 1968.
24. G. M. Hansler, Acting Assistant Regional Director of Environmental Control, Health, Education, and Welfare, tape-recorded interview with author, July, 1968.

9

Those Convenient
New Foods

S. newbrunswick is a serotype of *Salmonella* isolated from a baby chick in New Brunswick, New Jersey, in 1937. It had not received much attention because between 1947 and 1966, out of 50,782 salmonella isolations from human sources, only thirteen were the same as that identified from that tiny ball of yellow fluff.

The State of Michigan had never had a known case of human *S. newsbrunswick* until January, 1966, when two male infants—both less than six months of age—developed severe gastroenteritis. Even though they lived in different parts of the state and their families had never met, both had *S. newbrunswick* forms of salmonellosis.

The Michigan State Department of Health began an investigation. Both babies were fed the usual baby foods, and each had been on a formula made with widely distributed brands of instant nonfat dry milk. Inspectors from the Health Department purchased samples of dry milk at local retail stores, and the laboratory examined the products. No salmonellas were found.

Public health officials asked other state health departments whether they had any reported cases of *S. newbrunswick* illness. Seventeen did. In all, there were twenty-nine cases that had been reported across the country between March, 1965, and February, 1966. The product that carried the rare organism had to be something distributed nationally. But what?

One thing became quickly apparent when the cases were gathered together: 48 percent of the victims were infants less than a year old.

But what did the limited diet of infants have in common with the diets of adults? Solving the puzzle was made easier by the fact that one adult victim was a strict vegetarian, and thus a large number of dietary sources of salmonellas, such as chicken and cake mixes, could be eliminated.

The one thing that all the victims had in common was dried milk. Though multiple brands of milk were named, this did not discourage the researchers, since they knew that there was a frequent interchange of base powder among various producers of the final instantized product and that the majority of producers drew on the same large milksheds in the Midwest.

The trail finally led to a processing plant that produced an estimated 11 million pounds of nonfat dry milk in 1965, approximately 4 percent of the total instant nonfat dry milk in the nation. Some eight hundred farms routinely supplied the milk for the plant. A number of materials produced by the plant in question were found to be contaminated with *S. newbrunswick*; two of the employees involved in sifting and bagging operations were found to be infected with *S. newsbrunswick*; the air filter of the spray driers was contaminated with *S. newbrunswick*.

The mode of the first contamination of the plant was never identified. Health officials suspect that one of the eight hundred farms supplying the plant provided raw milk contaminated with *S. newbrunswick* and that the heat treatment in the plant was insufficient to kill the organism.

The case history of the outbreak, which occurred over a period of a year in seventeen states, was reported by Richard N. Collins, Michael D. Treger, James B. Goldsby, John R. Boring, III, Donald B. Coohon, and Robert N. Barr, in the *Journal of the American Medical Association*, March 4, 1968.[1]

The case illustrates two points:

1. THAT WE DEMAND CONVENIENCE—Most mothers have given up breast-feeding, and an increasing number are giving up mixing their babies' formulas in their own kitchens. Although there is little enforcement of standards in premixed formulas, and the possibility of contamination of dried milk is high, we are willing to take the chance, through ignorance or laziness.

2. THAT PUBLIC HEALTH OFFICIALS FIND IT DIFFICULT

TO RECOGNIZE AN EPIDEMIC IN CONVENIENCE FOODS. Outbreaks may occur over a long period of time and in widely separated areas; and because of faulty reporting, the source may not be discovered. The vast increase in new convenience products can be appreciated when we learn that of all the food products available for the housewife today, 66 percent did not exist ten years ago.[2] At the turn of the century, fewer than 100 food items were produced and processed by the food industry and they were items that consisted primarily of dried staples from nearby farms. Today, 8,000 to 10,000 food items are made available to the consumer, ranging from dried staples to ready-to-eat convenience foods.[2a] Today, more than 50 percent of the food on American dinner tables is processed.

According to a report of the Food Protection Committee of the National Academy of Sciences and National Research Council:

"The most outstanding development in the food industry has been the transfer of food preparation from the home to the factory. At the local level, food service organizations provide prepared meals and other food items on a large scale for industrial employees, schoolchildren and special groups. A wide variety of perishable prepared ready-to-eat foods is being distributed for home consumption by "carry-out" food services. In general, food production of this type is under purely local public health supervision.

"On the national level, large quantities of products are manufactured in individual plants for distribution throughout the country and internationally. In earlier years, such foods were primarily "sterilized" or "stabilized" canned or dehydrated products. However, the improvement in food distribution methods in recent years, particularly in applications of refrigeration, has made it possible for perishable products to be shipped over long distances and retailed in good condition."[3]

In 1940, the average housewife spent about five hours each day preparing three meals for her family of four. Currently, it takes her less than one and a half hours.[4]

Industry is currently investing more than $500 million yearly on new product developments that inherently involve a wide assortment of food safety problems. A panel on the development of public support for food protection programs at an American Health Association-sponsored conference in April, 1971, reported thus: "Unquestionably, these technological developments have provided improvements to the consumer as well as the food industry in the area of food production. However, too often such technological changes are equated with improvements in quality without discriminating between changes which could reduce health hazards and those introduced primarily for reasons of convenience or economic advantage. Coupled with the technological revolution in the food industry, environmental health and other regulatory agencies have been unable to keep pace with the overwhelming number of new problems that continue to rise."[4a]

An article by Dr. Abel Wolman, "Man and His Changing Environment," in the *Journal of Public Health*, November, 1961, pointed out: "The increase in the number and variety of processed foods is considered by some to be synonymous with a better and easier life. It seems to be assumed that most individuals for one reason or another are unwilling to accept foods in their natural state. Hence, every conceivable effort has been made to modify color, taste, structural appearance, texture, stability, nutritive value and substitute commodities. Such successful efforts to supplement, modify and conceal nature introduce innumerable problems of identification, of evaluation, of control and of assessment of long-term biological effects.

"To the orthodox problems of biological hazards in food, industry has added the major issues of chemical, biochemical, physical and other adjustments. The natural and artificially induced shifts in food habits, new methods of food technology, and the intentional and inadvertent use of additives find the health officer confronted with a baffling set of conditions unheard of even a decade ago."[5]

No one in America wants to go backward. We want both convenience and an abundant choice of food. Yet we want it, and should have it, free of hazard, and it should be of good quality. Too often, we are not getting what we expect and pay for. As proof, here are some examples that are just a fraction of such infractions on the market in 1971:

Peanut patties in 5 and 10 cent sizes. FDA inspection of manufacturer revealed "extensive rat infestation of plant and contamination of product samples with whole rodent excreta pellets," in Dallas, Texas.[5a]

Egg mix, for use in school lunch program in Memphis, Tennessee, and Kansas City, Missouri. Testing of the product in tin can containers showed salmonella contamination.[5b]

Vegetable juice cocktail, was found in San Francisco to contain lacquer flakes caused by a breakdown of the can's lining.[5c]

Breaded fish cutlets in Seattle, Washington, were found to contain decomposed fish.[5d]

Hash brown potatoes, prepared and packed under insanitary conditions in Little Rock, Arkansas. They were insect contaminated.[5e]

Others are:

KANSAS CITY DISTRICT—Dehydrated potatoes and frozen custard pies were seized in April, 1967, because of *E. coli*, excessive coliforms, and total bacteria.[6]

LOS ANGELES DISTRICT—Because of salmonella contamination, a pie company of Los Angeles, California, recalled more than 100,000 frozen-custard and pumpkin pies in April, 1967. Inspection of the firm that supplied liquid eggs to the bakery revealed inadequate pasteurization.[7]

ATLANTA DISTRICT—A Florida macaroni company and its president were fined a total of $12,000 on February 2, 1968, on charges of insect infestation.[8]

BALTIMORE DISTRICT—A railroad car of yellow shelled corn was seized at Greenville, North Dakota, on June 8, 1967, owing to an unsafe pesiticide residue, Captan.[9]

COLORADO DISTRICT—Prompt identification of salmonella by the Colorado Department of Health in a food-poisoning case resulted in a nationwide recall of frozen Chinese dinners manufactured in Minneapolis, Minnesota.[10]

At an American Public Health Association meeting on November 12, 1968, researchers from General Foods reported finding *Cl. botulinum*, Types A or B, in six samples of frozen vacuum pouchpack spinach in butter sauce. The samples were among one hundred randomly selected, commercially available cut green beans and frozen chopped spinach.

The authors of the report, N. F. Insalata, section head, General Foods Corporation, Post Division Research; J. S. Witzeman, J. H. Berman, and E. Borker, denied there was

an "eminent hazard" because if housewives followed instructions and boiled the pouch for fifteen minutes, the
botulinum would be destroyed.[11]

But what if they didn't follow instructions?

FROZEN FOODS

Former FDA Commissioner James Goddard said in an
interview with the author that the "major area of concern
of the Federal Food and Drug Administration in the
food-protection system is providing safe, good-quality frozen foods right from the time of manufacturing and processing into housewives' kitchens."[12]

A news release of the American Public Health Association put it in stronger terms in April, 1968: "Bargain day
for TV dinners may be a bad bargain for your stomach.

"A study, 'Incidence of *Clostridium perfringens* in Fish
and Fish Products,' by Dr. Mitsuru Nakamura and Kath-
erine D. Kelly, reveals that the incidence of *Clostridium
perfringens*, a bacterium which may cause acute gastroenteritis, is twice as high in TV dinners and potpies containing fish as in fresh or smoked fish."

Health Laboratory Science, a journal published quarter-
ly by the Laboratory Section of the American Public
Health Association, has featured a variety of studies of
food contamination in the past. This study, based on
selective sampling taken in Missoula, Montana, points out
that frozen foods are often subjected to thawing and
refreezing several times prior to sale. On sales days, when
the packages are stacked high and displayed under inadequate refrigeration, the bacteria present in the fish may
reproduce and initiate food poisoning when the food is
consumed.

The APHA article concluded: "The moral of the story?
Turn off your TV and go catch your own fish!"[13]

Actually, patents for freezing fish were issued to H.
Benjamin in England in 1842. Before that, cooling with
natural ice or in caves was practiced for centuries.

Mechanical refrigeration, developed in the late 1800's, allowed foods to be stored at about 40 degrees F., but frozen foods didn't really get started until 1940. Clarence Birdseye and others helped to develop quick-freezing processes, thus promoting frozen foods as a consumer item.[14] Annual consumption of frozen foods has risen from six pounds per capita, around World War II, to thirty or more pounds.

Because of their high water content, most foods freeze easily at temperatures between 25 and 32 degrees F. Although each food has a specific freezing point, quick freezing takes place in thirty minutes or less. If freezing is rapid, damage to the food is reversible. Slow freezing permits large, uneven crystals to build up; these puncture the food cells, and make the food mushy when it is defrosted.

According to *An Evaluation of Public Health Hazards from Microbiological Contamination of Foods*, a report of the Food Protection Committee of the Food and Nutrition Board, National Academy of Sciences and National Research Council, the primary hazard in frozen products lies in contamination and perhaps microbial growth or toxin production during preparation and processing and in failure to observe proper temperature control during storage, distribution, retailing, and especially in the home.

Frozen foods are very sensitive to temperature increases. In "Quality and Safety in Frozen Foods," published in the *Journal of the American Medical Association*, it was pointed out by Dr. Horace K. Burr and Dr. R. Paul Elliott, that most chemical reactions are 20 to 30 percent more rapid when the temperature rises 5 degrees F.; but certain deteriorative reactions in frozen foods may double, triple, or even quadruple their rate with such a temperature rise. They said that deterioration begins even at 0 degrees F, but that, if packaged suitably, frozen foods retain high quality for many months.[15]

Under present regulatory conditions, the manufacturers themselves are trusted to make sure their products are not contaminated with food-poisoning bacteria. No federal law sets limits on the amount of food-poisoning bacteria that can be present in frozen foods, although the United States Army sets such standards for the food it buys. Generally, federal and state authorities don't step in until the contaminated food has reached the market.[16]

Though 50 percent of the live organisms are killed in frozen food in six months and 75 percent in one year, the low count does not mean that the food is safe. It may have been handled in a very unsanitary manner and may have food toxins and other contamination.[17]

At the Frozen Food Packers National Association meeting in Chicago, in 1964, former FDA Commissioner George P. Larrick said:

"Routine bacteriological controls are not sufficient to detect potentially pathogenic microorganisms which may be present in raw ingredients.

"Pesticide residues in frozen foods are a problem requiring unique controls where raw agricultural commodities of a perishable nature are purchased.

"An ever-present problem, both for the frozen-food packer and the FDA, is the care employed in adding just the right amount of direct food additives for which there are tolerance limitations. There is a basic assumption in the use of these that the food manufacturer will take the necessary precautions to avoid misuse.

"When our inspectors visit your plant—*about once every two years on the average*—they need to know how good your controls are so we can evaluate how you are carrying out your responsibility day in and day out."

Mr. Larrick concluded: "The extent to which some in industry adopt control methods is influenced by the attention given the problem by federal, state and local officials."[18]

When the FDA conducted a survey of frozen-food plants which covered sixty-three factories in eighteen states, some three thousand samples were collected and examined. The results led to the conclusion that "sanitary and operating practices in plants were considerably below the level desired" for this kind of food product.

Even if the commercially frozen food is not of the quality we would like it to be, for the most part it is safe. The really dangerous frozen food is that prepared by the independent small-time merchant who wants to freeze something he makes himself. Delicatessens and specialty stores do it, most of the time without proper equipment and without any inspection at all. In such cases, one man's frozen dish may truly be another man's poison.

CANNED FOODS

When Napoleon had his hand in his shirt, he may have been hiding the first can of food.

All eighteenth-century soldiers traveled in peril of their diets of putrid meat and other inferior food. In order to improve the condition of his fighting men, Napoleon offered a prize of 12,000 francs for the invention of a method of food preservation for his army. A confectioner, Nicolas Appert, observed that food cooked in sealed bottles remained unspoiled as long as the container remained unbroken. Appert won the prize in 1809. By 1820, the first commercial canning plants were operating in the United States, and by 1840 canning was common throughout the country.[19]

Considering the number of cans produced annually, the record of the American canning industry has been remarkably good. Commercially canned food may be stored almost indefinitely under proper conditions. Sealed cans are heated during the canning process to destroy undesirable organisms. Although acid foods may react eventually with the tin and iron in the can to cause off-flavors, such a taste is not an indication of spoilage. Canned foods should be stored in cool places to retard such reactions. There is a considerable loss of vitamins when the temperatures of canned vegetables and fruits are too high.[20]

FERMENTATION

Man has used microorganisms to aid him in the production and preservation of his food for a long time. Beer, cottage cheese, buttermilk, sauerkraut, bread, and wine are all examples. Fermentation is a process of decomposition of the carbohydrates in foods. It is easily distinguishable from putrefaction, which results from the action of microorganisms on protein materials.[21]

DRYING

Drying food to preserve it is as old as man's discovery that the sun could stop decay.

In the 1700's, people put their food near the fire to hasten drying it or else spread it in thin layers in direct sunlight to hasten the process. In 1795, two Frenchmen named Masson and Challet made an important improvement in the process of drying when they built a dehydrator in which air heated to about 105 degrees F. was blown over thinly sliced vegetables.

The first dehydrated instant foods, however, are credited to the United States Army Quartermaster Corps. In World War II, the Corps came up with instant mashed potatoes that tasted very much like wallpaper paste. During the war the Quartermaster Corps issued a contract to industry to improve dried potatoes. The improvement was made, and a host of other dehydrated products were developed, such as eggs, milk, and some vegetables. The processing reduced shipping weight, eased storage problems, simplified distribution, and enabled soldiers to carry their rations with less effort in the field. Though these World War II products still lacked quality, and were difficult to rehydrate, they were the forerunners of today's many dried convenience products.

Prior to 1950, most dehydration was in cabinets or tunnels with forced blasts of air at high temperatures to remove water. Such methods caused damage to the product. Today, vacuum drying in cabinets is used for prune flakes, apple powder, and potato granules. Spray-drying and drum-drying are also used for dehydration. Improvements in heating methods, water removal, and temperature control makes spray-drying and drum-drying techniques applicable to more foods in greater volume. Coffee, tea, soup mixes, fruit solids and vegetable purées may be spray-dried.

A fairly new method is puff-drying. The product is partially dehydrated by circulated hot air; then it is exploded (puffed) by a gun. The food is finish-dried in heated bins to withdraw the last traces of moisture. By using this method, the United States Department of Agriculture has puff-dried carrots, beets, blueberries, and apple pieces.

Studies on a significant number of samples of various dehydrated fruits and vegetables have shown very few, if any, of them to be sterile. Very little information is available concerning the significance of microbial populations found in these products. [22]

A combination of both the freezing and dehydrating techniques is "freeze-drying." This is a process in which moisture is removed from the food while it is in a frozen state. The advantage of freeze-drying is that the water is removed in a manner that does the least physical damage to the food itself. Freeze-dried foods now include soups, casserole dishes, fruits for cereals and coffee. They retain their flavor, aroma, and stability much better than air-dried foods.[23] One critical problem with freeze-dried foods is that they absorb water more rapidly than other dried foods. Therefore they require special packaging and handling. These facts, plus the higher cost for water removal, mean that the cost of freeze-dried products is usually high.

A new process using microwaves to hasten water removal may have good possibilities. When the microwaves pass through the food, the molecules in the food attempt to align themselves with the electrical field, creating friction and heat. As the microwaves uniformly pass through the food, water is removed by the heat generated within the food.

Concentrated, dry frozen baby food has been tested by the University of Missouri School of Medicine. Reporting their findings in the *Journal of Diseases of Children*, the researchers said that forty-nine infants were fed with the products. None developed signs of food poisoning or allergy. The researchers also said that, without exception, "the mothers stated that this new baby food appeared fresher and more appetizing." The processing included the application of heat, reduction of moisture, and freezing. No colors, flavors, preservatives, or fortifying nutrients were used in the process.

The Wisconsin Alumni Research Foundation reported that studies of the frozen, concentrated baby food showed that the nutrients were effectively retained in processing and storage at freezing temperature. Bacteriological studies at the University of Chicago showed that no food-poisoning organisms were found and that the food could be safely stored in the deep freeze for about one year and in the refrigerator for about one month.[24]

MILDLY PROCESSED FOODS

Perhaps the most potentially dangerous of all convenience foods are the so-called "mildly processed."

Here is a quote from *An Evaluation of Public Health Hazards from Microbiological Contamination of Foods* by the Food Protection Committee of the Food and Nutrition Board, National Academy of Sciences and National Research Council:

"A variety of products in which microbial populations have been reduced in number by some mild bactericidal treatment—usually heat—is on the market. The final product is most commonly packaged in a metal can or a plastic bag, often under vacuum, and should be stored under refrigeration. Since these are not sterile products, there is an obvious danger associated with bad handling of the food either before or after processing. A few retailers still handle all canned goods as though they are sterile; consequently it is possible to see canned nonsterile products such as hams and bacon stored out of the refrigerated area.

"There are reasons for believing that the bacterial hazards associated with mildly processed products are greater than from corresponding unprocessed items. The principal objective of mild processing with most foodstuffs is to eliminate low-temperature spoilage micro-organisms which are in general highly sensitive to heating, smoking and irradiation. In untreated food, the normal flora serves two functions that concern the consumer; it quickly renders the food undesirable when storage conditions are poor, and in some cases it apparently competitively suppresses the growth of food-poisoning organisms. The former serves to warn the consumer of a potential danger. In pasteurized foods, the microbial balance is upset; the organisms that normally grow most vigorously on the stored foods are eliminated and conditions are probably made more favorable for growth and, perhaps, toxin production by potentially pathogenic organisms.

"If these pasteurized foods are packaged in vacuumized containers, an additional factor of risk exists where good refrigeration practice is not followed. The anaerobic conditions created are quite inhibitory to many of the surviving spoilage bacteria but may be suitable for the development of botulism."

The evaluation also said that the number of foods falling into this mild-processing category are increasing all the time, and already include poultry and meat products and party dips of various kinds:

"In many such processes, the margin between a microbiologically satisfactory product and a potentially dangerous one is clearly very narrow, and its maintenance is almost entirely dependent upon proper control of storage and transportation conditions."[25]

IRRADIATION

Irradiation of food would seem to be the answer to all problems of foodborne disease and quality loss. The United States Atomic Energy Commission, Division of Technical Information, in a brochure describing "Food Preservation by Irradiation," said:

"Oranges as sweet and juicy as if just picked—strawberries firm and free of mold—yet all many miles and weeks away from harvest. How is it possible? They've been radiation-'pasteurized'—exposed to low levels of nuclear radiation.

"Bacon, ham and chicken that haven't been near a refrigerator in months are fed to hundreds of soldiers who find them the equal of normal fresh meat. How? They've been radiation-'sterilized'—exposed to levels of radiation high enough to kill bacteria."[26]

Other than canning, radiation processing is the only original—that is nonnatural—method of preserving food. The amount of radiation to be delivered depends upon the food itself and the result desired. If the goal is prolongation of shelf life, or storage time, a pasteurization dose, generally from 200,000 to 500,000 rads, is sufficient. If the food is to be sterilized for long-term storage without refrigeration, the required dose is between 2,000,000 and 4,500,000 rads.

Even lower doses of radiation—4,000 to 10,000 rads—stop potatoes or onions from sprouting. Twenty thousand to fifty thousand rads kills insects infesting food. Doses of 200,000 to 400,000 are highly effective in extending the refrigerated storage of fresh fish up to thirty days. Bacon and other pork products, chicken and beef can be packaged and then irradiated at 4.5 million rads, resulting in sufficient preservation to permit a year's storage at room temperature.

Irradiation could save us money by stopping food spoilage, help underdeveloped countries without adequate refrigeration and food distribution, and generally be the answer to worldwide needs.

On February 8, 1963, the FDA ruled that bacon preserved by radiation is safe and fit for unlimited human consumption. Later, the government agency approved radiation to inhibit sprouting in white potatoes and disinfestation of wheat and wheat flour in this country. Such foods, however, are not for sale in the United States and chances are they may not be for a long, long time.[27]

In 1968, the FDA rejected the Atomic Energy Commission and Army's petition for clearance of canned irradiated ham. In addition, FDA officials said that all previous clearances of foods preserved by irradiation will have to be reevaluated to make sure the data justify the approvals. In a letter to Army Brigadier General F. J. Gerace, Commander of the Army's Natick Laboratories, the then

FDA Commissioner, Dr. James Goddard, said, "Our evaluation of these data [for use of gamma radiation in preserving ham] fails to establish that the proposed use of gamma radiation will be safe."

In a letter to AEC Chairman Glenn T. Seaborg, Dr. Goddard said he was "most disappointed" with Army data, and would have to have better data before new clearances are given. In discussing reasons for the denial of the petition, Dr. Goddard said that reported adverse effects on animal reproductive processes are "highly unlikely" to be due to chance. The FDA head cited a University of California at Los Angeles study on rats fed irradiated bacon and fruit compote which revealed that "the feeding of a diet containing irradiated bacon on the 1x level was associated with the reduction of 22 percent in weaned progeny and a 23 percent reduction in live-born progeny." Other adverse effects included increased mortality rates in rats fed a diet of bacon irradiated at the 2x level. "There was consistently reported a slight but persistent depression of body weight observed both in dogs and mice that were fed irradiated diets," according to the UCLA study.

Dr. Goddard also expressed concern over the tumor-causing potential of irradiated foods. Rats fed a diet of irradiated bacon and fruit developed more tumors than a control group of animals. "One study of rats reported three carcinomas of the pituitary gland among 52 animals on irradiated pork diets versus none in the 26 animals on the control diet. Since this is a rarely occurring type of tumor, this could be very significant."

A study in the 1960's at Cornell University showed that eating irradiated sugar can produce the same results as irradiation directly applied to the cell. The scientists experimented with carrot tissue and coconut milk . . . both high in natural sugars. They bombarded them with cobalt 60, causing radiation-induced cell mutations in both foods. Moreover, the chemicals produced by sugar breakdown in the foods were seen in transfer radiation effects into the cells of fruit flies, resulting in stunted growth and chromosome damage. All living cells contain sugar, the report emphasized, and human beings may suffer similar consequences from long-term consumption of irradiated food.[28]

There are also many studies that maintain that irradiated food is wholesome. In the *FDA Papers,* May, 1967,

Robert S. Roe, Associate Director of the Bureau of Science, summed it up:

"FDA scientists want to know if ionizing radiation causes radioactivity in the food, whether they produce toxic substances in the food or whether proposed treatments adversely affect the nutritive values of exposed food and whether the proposed treatments are effective for intended purposes.

"It is important that the greatest care be exercised, particularly at this stage of development, to insure that any regulations issued are scientifically sound and will, in fact, provide for safe and effective uses."[29]

Scientists kept up a running battle in the pages of *Science,* the publication of the American Association for the Advancement of Science, on the merits and dangers of food irradiation. The final consensus of opinion was that the burden of proof is on those manufacturers who want to irradiate food.

Perhaps radiation will prove to be the answer; perhaps we shall go back to older methods. In fact, United States Department of Agriculture researchers reported at the 1968 meeting of the American Institutes of Biological Sciences in Urbana, Illinois, that ultraviolet light (the same as that of the sun) can kill 99 percent of the harmful bacteria in maple syrup. They said control of bacterial growth by ultraviolet light irradiation is "simple, inexpensive, leaves no harmful additives, yet is 100 percent effective."

Actually, little is known about the microbiology of many of the newer foods and food-manufacturing processes.

The National Academy of Sciences and National Research Council report said: "Even as it concerns frozen foods, on which a considerable proportion of applied bacteriological research effort has been concentrated recently, there are many unexplained factors.

"The development of bacteriological guidelines is of the greatest urgency in the case of convenience and mildly processed foods. While there are strong public pressures in some areas for the regulation of the production of these products by the imposition of legal bacterial standards, the scientific basis for such standards has not yet been provided."[30]

NATURAL FOODS

When Eve handed Adam the apple, it may have had a worm in it, but it didn't have any pesticides, wax coloring, or chemical fertilizer. We all know the consequences of taking a bite of the first apple, but what about the aftereffects of today's fruit?

The proof that people are worried about it is the phenomenal growth within the past year of health food stores across the country where people are paying from two to three times as much for an "unadulterated" apple as they would for one from the corner supermarket.

Perhaps you can reduce the amount of "poisons" in your food by buying products from a "natural" food store. The problem is that you are really fighting a losing battle. As the chapter on pesticides in this book relates, bug killers, sprayed from planes and even by hand, get into the streams and into the air. If you drink water and breathe air, you are absorbing pesticides. The same is true about other chemicals in the air and water.

Dr. Philip Handler, president of the National Academy of Sciences, which oversees much of the checking of food additives, said: "There is nothing sacrosanct in the composition of natural foods; their composition reflects not nature's design for human nutrition, but rather the chemical mixture appropriate to the functioning of propagation of the species of plants and animals which are our foods. If fortification with synthetic amino acids or vitamins is the cheapest way to assure adequate nutritional levels, it is folly to stand in the way."[30a]

So-called food faddists who have been promoting the use of natural food, additional vitamins, and restrictions on the use of additives have lived to see the day when many of their contentions have been proved in scientific laboratories.

At the National Conference on Cancer of the Colon and Rectum in San Diego, California, in January, 1971, a

British cancer specialist in a speech to nearly 1,000 delegates said that colorectal cancer could be a product of the modern Western diet of overrefined foods and confections.[30b]

Many natural food enthusiasts had been blaming cancer and other ills on the American diet. At the conference on cancer, Dr. Wendell Scott, clinical professor of radiology at Washington University, St. Louis, Missouri, stated:

"If the additive itself be without intrinsic nutritional value, but somehow necessary to assure that the full nutritional properties of the foodstuff reach the consumer, and there is no substitute, rigorous standards are required to establish what level of risk is acceptable. Lethality is quite out of the question, but an incidence of one reversible untoward incident per 500,000 consumers, for example, may be an acceptable risk.

"If, however, the additive offers no form of nutritional value but is present to enhance flavor or texture, the public is entitled to assurance that such addition is quite without hazard, viz., failure to detect metabolic or physiological alterations, abnormalities of development, mutations, or neoplasia. By failure is meant that the odds against such unwanted effects are at least greater than a million to one at ordinary levels of intake, with no detectable chance of lethality whatsoever. We sadly lack a data base for making such statements concerning most additives or many drugs. . . .

"This is not to urge that food additives be removed immediately from all food preparations but, rather, that we embark upon the extremely large effort required to obtain such data. . . ."

Food at the natural food stores does taste better. Nevertheless, Americans should not have to pay three times as much to obtain food without unwanted chemicals as they would if they bought food from the corner grocery or supermarket. Most chemicals added to our food are for the benefit of the seller, not the buyer.

THE SUPER-SUPERMARKET

When you go into a supermarket, what do you look for? The items on your grocery list? The special bargains for the day? Few of us look for signs of good sanitation in any of the nation's more than 28,000 modern food stores.

The average supermarket handles 7,000 to 8,000 items. Most invest $100,000 for temperature and humidity control to keep food in top condition.[31] Yet physical damage to food products from breakage, bruising, and other mishaps in warehousing, delivery to retail stores, and in the stores is estimated at more than $50 million annually. About three-fourths of this damage occurs in the store, with employees causing about 60 percent of it and customers about 40 percent. Some of the damage is not recognized until after the customer gets the product home.[32]

Most women do not select their supermarket; the market selects them by establishing a store in their neighborhood. Fortunately, the majority of stores do a good job in food protection; but all have a problem in maintaining quality, and a few harbor sickness and even death on their shelves.

Inspection of supermarkets, like inspection of restaurants, food-processing plants, caterers, and farms, is spotty and sometimes non-existent.

I went into a brand-new, luxurious super-supermarket with Dr. Paul LaChance, Associate Professor of Nutritional Physiology at Rutgers University, New Brunswick, New Jersey.[33] There were wide aisles with thousands of products invitingly displayed. It looked like a housewives' shopping heaven to me. Yet he showed me fish in broken plastic bags inadequately protected and refrigerated; chicken in packages that were perforated, with legs sticking out and juice dripping into the cabinet. (Women shoppers were handling the chicken with their bare hands, pushing aside the packages.) Sliced bologna, inspected by a local health department, was two months old; honey jars were leaking; hams were unrefrigerated; potatoes had heart rot; meat was packaged so that the underside and sides were hidden from the shopper; processed foods were piled high; and some, such as dough and cheese pirogis, were soft instead of frozen.

Dr. George Kupchik, Director of Environmental Health, American Public Health Association, said there is little doubt that some supermarkets sell over-age milk and do not handle meats and seafood properly.[84]

Dr. LaChance could read some of the mysterious codes on packages to determine their age.

"See this?" he said. "Women believe the stores put the older products in front and the new products in back. They usually reach toward the back to select a package."

The product that Dr. LaChance had selected from the back was much older, according to the code, than the product in the front.

A letter published in *Consumer Reports*, August, 1967, said:

"I've noted that you occasionally advocate dating of perishable food products. I don't think you go far enough.

"Last year I worked for a convenience-type grocery chain, and I know that people unknowingly purchased milk that was one and two days out of date, sandwiches that were a week out of date, and cartons of orange juice and lemonade that were two weeks out of date. Almost always a fresh package was available on the same shelf.

"I well remember one day when a young mother purchased a few cans and jars of baby food. I noted an old code on a jar of prunes and a can of meat and vegetables, and after she left I figured out how old those items were. Both had been on the shelves since the store first opened 32 months before. Perhaps the contents were perfectly safe, and good, but I lost my appetite for supper that day just thinking about it. . . .

"Most products are dated in code. I know that you cannot possibly explain all of the hundreds of systems used. But why don't you recommend a requirement that every grocer prominently display an explanation of all date codes used in his store?

"Orlando, Fla."

"F. C. W."

The editors of *Consumer Reports* answered: "Better yet, why don't food packers mark the labels with a clearly legible date indicating the reasonable shelf life of their products? Then the cake mixes wouldn't sit overlong on the shelf and the customers would have known whether those baby foods were perfectly all right after 32 months in the store."[35]

Today, good store management is really up to the owner's skill and conscience.

The career of Dave Fern, president of 230 Wakefern Food markets and honorary chairman of 78 Supermarkets General, spans the era from the time of the small neighborhood market to that of the giant chains of supermarkets. He began working as a grocery clerk in 1912.

Mr. Fern, who sold Dave's Supermarket in Milburn, New Jersey, a number of years ago, still works there almost every day. "I like to be with people . . . give them some of the personal touch we had years ago," he said. Dave's Pathmark Supermarket now handles five thousand cases of products a week. Trucks pull up to the door at 6:30 A.M. and are unloaded automatically. The perishables are placed in a refrigerator. The nonperishables are stored off the floor on shelves. An exterminator services the store once a week. The Board of Health inspects at least once a week.[36]

Dr. LaChance and Weems Clevenger, District Director of the Food and Drug Administration, New York, are both food quality experts.

Dr. LaChance and Mr. Clevenger and other food experts with whom I spoke agreed that housewives and other shoppers who see a breach of sanitation in the store or who find they have spoiled food once they get home with their packages should report this to the store manager. If the situation is not rectified, then the customer should alert the local board of health.

Most store managers want to please their customers, and will go to great lengths to do so. Mr. Fern, for instance, said he had a refrigerated unit to hold produce.

"I had it in here for six months and then got rid of it because the men weren't changing the produce often enough. They left it in the unit too long. Now they have to carry all the produce to the refrigerator at night. In the morning, they must look it over and see if it is still right up to par. If not, we sell it for a few pennies. Actually, we would give it away, but the women wouldn't take it under those conditions. They would be suspicious."

Supermarket employees, in general, are paid better than many restaurant employees. At Dave's, the employees are unionized, and earn an average of $110 a week. Butchers earn as much as $150 a week. For Dave's and the other supermarkets in the chain there is an employee training program. Every department has a department manager, and there is an overall manager for the store.

Probably, the area that is weakest in all supermarkets is that of frozen foods. Most states do not have a frozen-food code that makes it unlawful to handle frozen foods at temperatures above 0 degrees F. But almost every item in a supermarket will deteriorate in time, regardless of whether it is merchandised in fresh or processed form.

Here are some hints on protecting your family's health when shopping in a supermarket:

SANITATION

Is the store clean? Are the floors dirty? Are the shelves messy?

Are the people who work in the store untidy-looking?

Are there bathrooms available for customers? For the help? Are the bathrooms clean and properly equipped with soap and paper towels?

REFRIGERATION

Meat is supposed to be held at 28 to 38 degrees F.; produce and nonfrozen dairy products at 35 to 45 degrees; ice cream at −12 degrees.

Are hams and other processed foods that require refrigeration under refrigeration? How about eggs? They should be refrigerated too.

FROZEN FOODS

Check the condition and temperature of packages. Are they in cardboard cartons that interfere with cold air?

Are the products above the load line in the cases?

Does the thermometer in the box register 0 degrees or below? Remember, the thermometer is usually in the coldest spot in the case, so there should be a good margin of safety.

Is there evidence of leakage about the package of frozen food? Your purchases should be solid, with no material adhering to the outside of the package.

If large quantities of ice crystals are present on the inside walls of a package of frozen fruit or if the fruit is somewhat shriveled, it is suggestive of thawing and subsequent refreezing or longtime storage. If there is considerable ice on the bottom of the package of dry-packed fruit, the item may have been thawed and refrozen.

PRECOOKED FROZEN FOODS

Appearance and odor may help to tell you, once it is defrosted, that it is not of good quality. It will indicate that the product has fermented during processing by reason of a slow freeze or that it has been allowed to thaw.

A dry, bleached surface is indicative of improper packaging, excessive storage periods, or widely varying temperatures in frozen storage.

CANNED FOODS

Don't look for bargains. Never buy an unlabeled or a dented can.

Acid foods may react eventually with tin and iron in the can to cause an off-flavor. This undesirable flavor is not an indication of spoilage. However, it is better to be safe than sorry. The deadly botulinus sometimes causes off-odors and off-tastes. Canned food should be stored in a cool place to retard off-flavors. There is also a considerable loss of vitamins when the temperature of canned vegetables and fruit is too high.

JARS

Never buy a jar that is sticky or leaking.

PRODUCE

Check the bottom of the lettuce. If it has a brown instead of white ring, it is old, and may be an indication of the practices of the management.

THINK TWICE ABOUT
BUYING SOMETHING NEW

According to Jean Judge, food-marketing specialist at Rutgers College of Agriculture and Environmental Science, "The customers who typically buy new products are usually those who most frequently reject the traditional housewife's role in food purchasing and preparation and thereby set the buying patterns for new products.

"They are most likely to be those who live in the urban areas of the Far West or Northeast, are well educated, represent families wherein the housewife is employed outside the home, have medium high incomes. Typically, families with young or school-age children are more likely to use new products, especially convenience items, than families without children."

Manufacturers of products know that an item is not profitable for more than two to five years, and for some, even less. Yet, the word "new" does not necessarily mean better.

Consumers Union found that a new food package which allowed the consumer to put a package of bacon into the toaster for instant cooking could cause a severe shock.[37]

Smoked fish that was packaged in vacuum-sealed envelopes turned out to be a wonderful medium for the growth of botulinus, and a score of persons died as a result.

This does not mean that food processors and vendors should stop trying to make improvements and developing conveniences for Americans. It does mean that each "new" product should be thoroughly tested for safety first. Like conservative doctors prescribing medicine, we should not rush out to buy something new, but hesitate long enough to see if it is really of good quality, and safe.

We should demand adequate inspection of food processors and supermarkets and other vendors of food.

Our best protection, however, remains our own skill and education as a buyer.

9 NOTES

1. Richard N. Collins, M.D.; Michael D. Treger, D.V.M.; James B. Goldsby, M.S.; John R. Boring, III, Ph.D.; Donald B. Coohon, D.V.M.; Robert N. Barr, M.D., "Interstate Outbreak of *Salmonella newbrunswick* Infection Traced to Powdered Milk," *Journal of the American Medical Association*, March 4, 1968, Vol. 203, No. 10.

2. *An Evaluation of Public Health Hazards from Microbiological Contamination of Foods*, National Academy of Sciences and National Research Council Publication 1195, 1964.

2a. National Conference on Food Protection, Denver, Colorado, April 4–8, 1971.

3. *The Sciences*, New York Academy of Sciences publication, May, 1965, Vol. 4, No. 12.

4. *Food Facts from Rutgers*, September, 1967.

4a. National Conference on Food Protection, *op. cit.*

5. Dr. Abel Wolman, Professor Emeritus of Sanitary Engineering, Johns Hopkins University, "Man and His Changing Environment," *Journal of Public Health*, November, 1961.

5a. *FDA Weekly Recall Report*, January 7–13, 1971.

5b. *Ibid.*

5c. *Ibid.*, April 8–14, 1971.

5d. *Ibid.*, March 25–31, 1971.

5e. *FDA Papers*, April, 1971.

6. *FDA Papers*, July–August, 1967.

7. *Ibid.*

8. *FDA Papers*, April, 1968.

9. *Ibid.*, October, 1967.

10. *Ibid.*, June, 1967.

11. N. F. Insalata, J. S. Witzeman, J. H. Berman, and E. Borker, "The Incidence of *Clostridium Botulinum* in Frozen Vacuum Pouch-Pack Vegetables," meeting of the American Public Health Association, November 12, 1968, Detroit.

12. Author's tape-recorded interview with Dr. James Goddard, New Brunswick, N.J., 1968.

13. American Public Health news release, April, 1968.

14. Grace M. Urrows, "Food Preservation by Irradiation," U.S. Atomic Energy Commission, Division of Technical Information, April, 1968.

15. Horace K. Burr, Ph.D., and R. Paul Elliott, M.D., Albany,

California, Council on Foods and Nutrition, "Quality and Safety in Frozen Foods," *Journal of the American Medical Association*, October 29, 1960, Vol. 174, No. 9, pp. 1178–1180.

16. William Dorsey, Chief of Richmond, Va., Public Health Laboratories, "Health Bulletin," June 6, 1964.

17. "Frozen Fruits, Vegetables and Precooked Frozen Foods," *Recommended Methods for the Microbiological Examination of Foods*, ed. J. M. Sharf (American Public Health Association, 1967), p. 97.

18. George P. Larrick, former Commissioner, U.S. Food and Drug Administration, before National Association of Frozen Food Packers, Chicago, Ill., March 20, 1964.

19. Grace M. Urrows, *op. cit.*

20. *Let's Talk About Food*, ed. Philip White (American Medical Association, 1967), p. 124.

21. Grace M. Urrows, *op. cit.*

22. *Recommended Methods for the Microbiological Examination of Foods*, p. 65.

23. *Food Facts from Rutgers*, January–February, 1968.

24. Gerard Van Leeuwen, M.D.; Emily J. Guyer, M.D.; University of Chicago, "New Concentrated Dry Frozen Baby Food," *American Journal of Diseases of Children*, January, 1961, Vol. 101, pp. 18–22.

25. *An Evaluation of Public Health Hazards, loc. cit.*

26. Grace M. Urrows, *op. cit.*

27. Reports in the press, July and August, 1968.

28. Dr. Richard D. Holsten, Cornell University doctoral thesis, 1965.

29. Robert S. Roe, *FDA Papers*, May, 1967.

30. *An Evaluation of Public Health Hazards . . . , loc. cit.*

30a. Dr. Philip Handler, "Can Man Shape His Future?" speech, before the Third International Congress of Food Science and Technology, Washington D.C., August 11, 1970.

30b. "Is Roughage the Key to Colon Cancer?" *Medical World News*, January 29, 1971.

31. Kermit Bird, Agricultural Economist, Marketing Economic Division, Economic Research Service, "Ketchup for a Small Boy: The Marketing Marvel," *The Yearbook of Agriculture*, 1966.

32. R. W. Hoecker, Chief of the Wholesaling and Retailing Research Branch, Transportation and Facilities Research Division, Agricultural Research Service, "Supermarket Safeguards," *The Yearbook of Agriculture*, 1966.

33. March, 1968.

34. Dr. George Kupchik, Director of Environmental Health, American Public Health Association, tape-recorded interview with author, February 26, 1968.

35. *Consumer Reports*, August, 1967, p. 454.

36. Interview with author, July, 1968.

37. *Consumer Reports*, February, 1967, p. 64.

10

Some Questions and Recommendations

It is not the purpose of this book to be sensational, nor is it to attack farmers, food processors, restaurants, government agencies, or any specific group. Its aim is not to point out that illness and death have resulted from what we eat, although this is certainly true. The author has no desire to hinder technology in its development of new and better means of preparing, packaging, and serving food.

The goal of this book is to raise questions.

Have we, like unsupervised children with beginners' chemistry sets, unrestrainedly poisoned ourselves and perhaps our descendants?

Have we allowed our technology to surpass our biological safeguards?

In a report to the Secretary of Health, Education, and Welfare by the Task Force on Environmental Health and Related Problems, dated June, 1967, it was pointed out:

"We know something of air pollution, but we know little about the potential hazard of 500,000 to 600,000 synthetic chemicals and other compounds on the market today. We know something of water quality but little of the effects of trace metals. . . .

"An individually acceptable amount of water pollution added to a tolerable amount of air pollution, added to a bearable amount of noise and congestion can produce a totally unacceptable health environment. . . .

"It is entirely possible that the biological effects of these environmental hazards, some of which reach man slowly and silently over decades or generations, will first begin to

reveal themselves only after their impact has become irreversible.

"Not only have we overwhelmed many of nature's processes for environmental stability; we have misused, without knowing it, biological processes upon which the preservation of life depends. By allowing tiny amounts of pesticides to enter our waters, we have set in motion processes that can lead to the destruction of birds that feed on fish that feed on plants that draw the pesticides from the water. Our ignorance of the consequences of our deeds may be innocent, but it is ignorance we can no longer tolerate."[1]

Ironically, with all our supermethods of growth, processing, distribution, and variety, we Americans are not eating as well as we did only a decade ago. A survey released by Secretary of Agriculture Orville Freeman in 1968 showed that 21 percent of our households had poor diets compared with 15 percent in 1955. The ability to pay was not the criterion. One in every ten families with incomes of $10,000 and over had poor diets. The Department of Agriculture also found that families were consuming more soft drinks, instead of milk and milk products, and eating less vegetables and fruits.

A comprehensive survey of 2,500 studies of nutrition, in 1969, showed that the most poorly nourished in America were infants, including many from upper income families. Most were not consuming sufficient quantities of calcium, vitamin A, vitamin C, iron, and iodine.[1a]

In 1970, Robert B. Choate, Jr., a citizen lobbyist on the issue of hunger, raised a tempest in a cereal bowl when he testified before a Senate committee that 40 of the leading dry breakfast cereals were so low in nutritional content that they constituted "empty calories."[1b]

The cereal industry's paid consultants in the universities hastened to defend the breakfast food by saying that dry cereals are eaten with nutritious milk.

As unbelievable as it may seem, we do not have accurate methods for determining what happens to the vitamins in our foods from the time they are harvested or processed until the time we eat them.[2]

We cannot answer the question of whether severe heat and dehydration processes in the preparation of precooked and easy-to-prepare "convenience" foods affect the nutritional quality of our food.[3]

In 1970, the Department of Agriculture announced that researchers had found that milling and bleaching of wheat for bread and cakes destroys 90 percent of the vitamin E, and 85 percent of vitamin B_6.[3a]

In 1971, Dr. N. R. Di Luzio, of the Tulane University School of Medicine, reported that indications of toxic substances in the blood of 78 persons out of a sample of 81 suggest that either these persons are eating too much "polyunsaturated fats" or not enough vitamin E.

Vitamin E is the body's natural protector against peroxidized or rancid fat. It is necessary to keep blood, arteries, and heart in good working order. In recent years, the American Heart Association and many physicians have recommended polyunsaturated fats in the diet to lower cholesterol, a substance that clogs arteries. Newer research has shown that polyunsaturated fats deplete the body of vitamin E, which is necessary to protect the heart and arteries. [3b 3c]

This is just one of the uncertain areas in nutrition research.

We do not know the effect of a pregnant woman's diet on the fetus. Previously, scientists believed that the placenta, which provides nutrition for the growing embryo, protected it from harmful chemicals in the mother's system. Now it has been discovered that not only does the placenta not protect the child; it may even concentrate and increase the harmful substance delivered to the fetus.[4]

Infectious hepatitis appears to be directly related to contaminated drinking water, but how does the virus get into the water and how can it be removed or neutralized? No one knows![5]

The Food and Drug Administration has estimated that the American people are being exposed to some 500,000 different substances, many of them over very long periods of time. Yet, fewer than 10 percent of these substances have been catalogued in a manner that might provide the basis for determining their effects on man and his environment.[6]

For instance, do pesticides cause cancer? Rutgers University reported in 1968 that certain azo (nitrogen) compounds are formed as the result of microbial action in soil treated with pesticide. It has been proved that some azo compounds cause cancer in experimental animals; other compounds are suspect.[7]

Are there significant amounts of pesticides still in our food when we eat it?

Researchers from Washington and California state that commercial and home washing methods cannot remove parathion residues from vegetables; in fact, home methods of preparation may increase the level of residues.[8]

Can natural foods cause cancer? Increasing evidence that the cycad plant causes malignancy is being gathered. How many other plants that we eat may also cause cancer?[9]

What are the effects on the heart of nitrates and nitrites that we ingest with our food and water? Scores of deaths that have been proved to be from these chemicals are mentioned in the scientific literature; yet we continue to allow the deliberate addition of such substrates to food for babies.[10]

Questions about the effects on our health of what we eat and drink are almost endless. No one knows—and there are no figures available on—the true incidence of food poisoning in the United States because so many incidents are unreported or because deaths from food poisoning have been wrongfully attributed to other causes. Scientists at the National Center for Urban and Industrial Health, Cincinnati, Ohio, point out that more than half of the two hundred outbreaks of food poisoning reported annually in this country are caused by organisms that are not—but often could be—identified.[11]

Bernard Aserkoff, M.D., Chief of Salmonellosis Unit, National Communicable Disease Center, Department of Health, Education, and Welfare, Atlanta, Georgia, said in a letter to the author: "It is important to realize that this total [100,390] represents a small fraction of the number of cases of Salmonellosis that actually occur. Of the total cases each year, only a small percentage are ill enough to report to a physician. Of those who seek medical assistance, only a few have stool cultures taken to document the cause of their illness. We at the CDC estimate that only approximately 1 percent of all cases of Salmonellosis are ever reported. In fact, the Salmonellosis problem has been compared to an iceberg of which we can see only a small portion of the total magnitude of the problem sticking up over the surface."[12]

How well are our government agencies protecting us from harmful food or food of poor quality?

The Task Force on Environmental Health summed it up for the Secretary of Health:

"The Task Force finds that many products available on the consumer market, including food and household equipment and appliances, are processed or manufactured in ways that render them hazardous to human beings under normal use, and that at present there is no adequate means of protecting the consumer from such products.

"In 1965, a total of 711 firms suspected of producing harmful or contaminated consumer products refused to let the FDA conduct inspections. Some 515 refused to furnish quality or quantity formulas to the Administration; 26 denied the Administration the opportunity to observe a manufacturing procedure and 153 refused FDA personnel permission to review control records.

"The consumer was the victim.

"Unlike most government regulatory agencies, the FDA does not have subpoena authority either to summon witnesses or to require firms to divulge pertinent records. It has requested this investigative authority to allow it to do a better job of protecting the American public. To this date, the request has been denied [by Congress].

"With respect to foods, the Department must have considerably more authority to assure adequate protection of the public health. The Department must be able to evaluate the synergistic effects of food additives so that the consumer is protected from threats that cannot be detected by the separate analysis of the individual food additives. This means evaluating the sum total of the toxicological effects of a mixture.

"Furthermore, the Department must have, as it does not now, adequate authority to inspect and evaluate the processing of foods to make certain that their safety is not impaired through the effects of a process which may or may not involve the use of additives."[13]

In my tours of food-processing, handling, and retailing establishments, I learned that federal and local inspectors first sign in with the management. A health officer explained to me: "We do it out of courtesy. Most people have nothing to hide and if they do, they can do it so well, the notification of our presence wouldn't make one bit of difference."

The shortage of qualified personnel for the purpose of food protection, ranging from physicians to local inspec-

tors, is acute. For instance, the FDA is desperately in need of "qualified specialists to review and evaluate scientific and clinical data." Unfortunately, our government is offering these physicians $17,953, when such persons could make from $20,000 to $100,000 in private practice or industry.

Some states have more positions open for sanitarians than they have positions filled. Again, pay is a factor. An Ohio city with a population of 80,000 advertised for a Category III sanitarian at a salary from $6,000 to $8,436. The requirements included a college degree, courses in sanitary and/or biological sciences, and experience in food-service vending, food establishments, and general sanitation.

A study of the characteristics of state food and drug employees showed that many have less than a high school education and that 41 percent of the state milk and dairy supervisors, 71 percent of food-inspection supervisors, and 84 percent of weights-and-measures supervisors have no degrees.[14]

This does not mean that they are not better at their jobs or sharper than many young college graduates with advanced degrees. However, inspectors without training in microbiological and chemical hazards, and who have inadequate salaries and are political appointees, are prime targets for corruption.

Just how prevalent corruption can be was demonstrated in New York when twenty-one milk firms, including most of the major milk distributors and suppliers in the metropolitan area, were indicted in 1968 for paying off sanitarians. One of the accused supervising sanitarians charged with graft had been on the job for more than forty years; another, for more than twenty-five.[15]

In *Public Health Hazards of Microbiological Contamination of Foods,* a report of the National Academy of Sciences and the National Research Council, it was pointed out:

"The food industries are highly competitive and tend to be concerned primarily with consumer acceptance of new products on the basis of appearance, flavor, texture and convenience. Their research and quality-control programs often do not delve deeply enough into the microbiological problems of food safety. For example, the incorporation of contaminated egg products into dry cake mixes and the

packaging of nonsterile products under pressure or in a vacuum may permit the multiplication of, and toxin production by, food-poisoning organisms such as *Clostridium botulinum* or *Staphylococcus aureus*.

"Government agencies have also been hard pressed to keep abreast of new developments because the resources for food protection programs have not kept pace with the rates of technological change or population growth. Their programs are largely based on visual inspection for gross defects in products, facilities and procedures where food is processed, stored or prepared for serving. These techniques are not adequate to cope with centralized manufacturing, mass distribution, mechanical dispensing and commercial catering of prepared products."[16]

Dr. Walter Modell, Professor of Pharmacology at Cornell Medical College, recommended in 1968 that the FDA be separated from the Health, Education, and Welfare Department, where it "suffers from exasperations, insecurities, anxieties and the oscillations of department whims, and set up as an independent commission, responsible directly to the President."

"The job of the FDA is complicated by crucial overlapping of functions with the Department of Agriculture. Thus, the FDA does not deal with pesticides and herbicides and other toxic materials that contaminate food, apparently because Agriculture can force the issue as it chooses. The FDA, therefore, has ruled that this very toxic group of drugs are not drugs.

"The FDA lacks prestige and has such a modest salary ladder it has great difficulty attracting promising, well-trained junior members, especially physicians."[17]

Dr. William C. Grater, head of the Drug Committee of the American Academy of Allergy, believes the American public has been badly served by the FDA, Congress, medical experts, and food processors. Sometimes, he said, the FDA has approved chemical agents despite inadequate testing, only later to withdraw them "on the basis of new studies." Some of the pressure has come from other countries. (Oil of calamus, used for years in fruit, chocolate, root beer, vanilla, and so on, was withdrawn in 1968 because it was found to cause malignant tumors in animals. NGDA, used in shortening and other such products, was removed from the "generally safe" list after Canada banned its use there.)[18]

"And there are no ways for the consumer to know what has been added," Dr. Grater said. "The law requires only that labels say artificial coloring, flavoring or preservative added. The label does not have to list them. If a person is sensitive to those things, there is no way for him to know what not to eat.

"I think the less chemicals you put in a person's environment, the better off he is, whether he eats them or breathes them or soaks in them," Dr. Grater concluded.[19]

Though Dr. James Goddard, when he was FDA commissioner, concentrated on increasing the margin of safety in drugs, he admitted that food additives did not even receive the testing that drugs did. He said that FDA laboratories did thirty-seven thousand assays a year but that, in order to do adequate testing on the important drugs alone, they should do a million individual analyses a year.[20] He also admitted that the FDA could not adequately inspect the eight thousand drug plants in America. What about the hundreds of thousands of food and cosmetic plants?

The FDA has been and still is understaffed, underfinanced, and demoralized. It was after three top officials of the agency were ousted in 1969, shortly after cyclamates were removed from products, that Dr. Charles C. Edwards, a management specialist from Chicago, took over the reins. Like his predecessors, Dr. Edward is a man of skill and good intentions, but the situation at the FDA has not improved. It may be that the agency is ungovernable.

George P. Larrick, who was the first of the four FDA commissioners in five years, said, in 1964: "The extent to which some in industry adopt control methods is influenced by the attention given to the problem by federal, state, and local officials."[21]

Although they are rare, there are criminals in the food business, from the shady buyers of contaminated foodstuff to truck hijackers to strong-arm vending-machine providers, and we are not being adequately protected from them.

Representative Benjamin Rosenthal, subcommittee chairman of the House Government Operations Committee, in a public hearing in March, 1968, offered proof that coffee, canned hams, and pre-cooked frozen dinners that the Army Supply Agency had classified as unfit for human consumption were being sold in supermarkets.[23] He did not learn this fact from an inspector but from one of his

constituents who had noted a label on a can of coffee that said "procured under U.S. Government Specification and for Military Issue and Sale to Authorized Commissary Patrons Only." Laboratory tests showed that the coffee contained "a harsh type of coffee not permitted by specification, and some coffee contained dust particles."

The subject of chemical additives in our food has perhaps received more attention than many other aspects of food safety. Yet, relatively little has been done about it.

Representative Joseph Minish (D–New Jersey) has tried unsuccessfully to get legislation passed to verify the safety of all food additives.

"While I am gratified that the GRAS list substance will be reviewed," Congressman Minish said, "I am still dissatisfied."

He pointed out that although the FDA was willing to consider all available evidence about GRAS substances, this places reliance solely on evidence already available—and in some cases would rely on lack of evidence.

"We cannot assume foods are safe," Mr. Minish said. "We must assure it."

"Since certain food additives, just as certain pesticides and other chemicals, have residual properties, the imposition of tolerance upon these substances will not eliminate their threat to health."[23]

According to Weems Clevenger, the District Director of the Federal Food and Drug Administration, New York, imports form one of the weakest links in the chain of food protection. Of the $27 billion total of annual United States imports, nearly a sixth—more than $4 billion—is accounted for by foods, drugs, and cosmetics. The FDA is responsible for the same safety, wholesomeness, and purity of these imports as it is for domestic products.[24]

Unlabeled canned goods and canned goods without English labels are referred by Customs to FDA inspectors. In one case, a Customs inspector became suspicious of activities in an enclosure at the far end of a pier, and reported them to an FDA inspector. Investigation disclosed that floor sweepings of coffee beans were being picked up there and packed for sale. The sweepings were contaminated by assorted filth from the dock.

According to the *FDA Papers* of April, 1968, of vital concern to Customs officials and the FDA are the relatively new Customs Bureau procedures known as "Imme-

diate Delivery," under which an importer can receive his shipment immediately upon arrival, and file the customs form within six days thereafter. The procedure is used at airports, seaports, and border ports, and is a boon to international traders, especially dealers in perishable commodities. But, without some means by which Customs can let FDA know about questionable merchandise, dubious goods may be in the hands of importers before FDA receives the necessary information.[25]

Mr. Weems Clevenger, New York District Director, said that FDA receives a list of imports, and selects the items to check. But he admitted that the FDA does not have sufficient manpower to do more than spot-check something its inspectors may think is hazardous. He recalled one questionable item, cheese with bits of salmon and bacon added. "When we asked the country of origin what inspection the cheese had received there," he related, "they replied that they smelled the cheese, and if it smelled all right, they passed it."

The rapid increase in containerization—the practice of shipping huge packages with many different items in each—is another new headache for food inspectors. For instance, when such containers arrive in the United States, should they be opened at the port of first arrival or be shipped to the inland point near the ultimate destination?

If this book raises many questions, it also presents some obvious answers. I asked Mr. Clevenger what he would do if he had unlimited power to improve the safety of our food supply:

"I would like to see more involvement of top management in industry," he said. "They spend more time dealing with money and promotion campaigns. We have been calling top executive officers of each drug firm now under intensified investigation, and laying out the whole sordid record before them.

"What happens in a food or drug plant is that an inspector goes in and gives the report to the plant manager. The plant manager doesn't want to be the bearer of bad tidings, especially when the boss holds him responsible. Many top executives do not know what is going on in their own plants."

From government experts like Mr. Clevenger, and from university scientists, physicians, and others concerned with

what we eat, here are recommendations for steps that should and could be taken without delay:

INCREASED KNOWLEDGE
ABOUT FOOD

We need research programs to produce new basic knowledge about both natural and processed foods in relation to health. This should include, but not be limited to, the study of new production, processing, and marketing methods as they may affect the public health.

We need more food science departments in institutions of higher learning in the nation. There are at present thirty-nine, and many of them are acutely understaffed and underfinanced. By 1974, the schools estimate a total capacity of granting 1,356 to 1,376 graduate degrees a year. As a report at an American Public Health Association conference on Food Protection pointed out in 1971: "In relation to the size of the food processing industry, as already indicated, the number of trained food scientists now graduating or whom one may expect to see graduating in the years ahead is woefully small, particularly when one considers that many graduates assume academic or governmental, rather than industrial positions."

Physicians need to study nutrition, both in medical school and afterward. Although it has recently been reported that heredity probably plays a part in how an individual reacts to food, medical schools present very little in the way of nutritional studies. Food illnesses and allergies are poorly diagnosed. According to Dr. George Briggs of the Department of Nutritional Science at the University of California, most nutrition studies are "quite worthless because of inadequate experimental design in regards to diet composition and control groups."

IMPROVEMENT IN THE DETECTION, INVESTIGATION, AND REPORTING OF FOODBORNE ILLNESSES

The case of the garlic-bread eater, paralleling the more recent Bon Vivant vichyssoise soup incident mentioned earlier, demonstrates the fatal flaw in our food protection system:[26]

On September 24, 1970, the Belmont Restaurant, of Arlington, Virginia, reported to the Virginia Department of Agriculture that they had purchased from the Skinker Specialty Food Co., of Alexandria, Virginia, a quantity of "Spice of Life Meat Tenderizer," which when applied to meat caused the meat to become a bright green, and on cooking, the meat turned a bright red.

The Virginia Department of Agriculture found, upon analysis, that the meat tenderizer was 97 percent sodium nitrite. This was brought to the attention of the FDA, and the agency issued a warning about the tenderizer, in November, 1970. The firm that manufactured Spice of Life and another brand, Country Taverns (both brands were almost pure sodium nitrite, but no one found this out because of mislabeling), voluntarily recalled the 1¼-ounce jars it had issued. Subsequently, the firm extended its recall to all 2¼-ounce containers as well. On November 13, 1970, the recalling firm and/or the firm's salesmen telephoned all known recipients of the 2¼-ounce jars, requesting that they promptly recall it from all their accounts, by any suitable means.

On November 13, 1970, a press release was issued informing the public of the mislabeling and potential danger.

There were no adequate records of where the bottles of tenderizer were subsequently sold; no adequate check on the jobbers who were supposed to notify their customers of the danger.

On March 14, 1971, at the Tap Inn, in Washington, D.C., a man ate a meal of halibut, topped with a cream and cheese sauce and garlic bread. The man died shortly

after. The cheese sauce and garlic bread were laced with Spice of Life meat tenderizer.

There was no public alarm sounded by the incident until March 19, when the *Washington Post* carried a story about the poisoning. A day later, the FDA issued a warning that was broadcast by radio and TV and printed in the evening papers.

Jars of Spice of Life and Country Taverns are still on kitchen shelves, as far as anyone knows.

We must insist on an effective early warning system when any food poisoning occurs, whether it is chemical, viral, or bacteriological.

Physicians, institutions, restaurants, and all food-handling establishments must be required by law to report suspected outbreaks.

Public health laboratories should have the personnel and equipment to detect specific organisms and chemicals causing foodborne illnesses.

PESTICIDES

A reevaluation of the entire use of pesticides should be undertaken, and the use of hydrocarbons such as DDT should be outlawed.

Pesticides should be colored for identification, and laws should be created to govern their packaging and shipment.

Nonchemical methods of pest control should be encouraged by means of greater financial support.

CHEMICALS

Nitrites, which are salts or esters of nitrous acid, should be banned from all products immediately. Production of baby foods containing them should be forbidden.

Immediate efforts should be made to survey the true amount of nitrates, which are salts or esters of nitric acid, in our water and food, and immediate steps should be taken to reduce such residues.

There is no doubt that nitrates and nitrites are harmful, as the pitiful sight of the graves of tiny children across the country will attest.[27]

Vitamin D should be restricted and its dangers publicized.

All chemical additives should be restudied by objective scientists—and that includes chemicals approved under the "grandfather clause." Just because something harmful may have been used by our ancestors is no reason it should poison us or our descendants.

All chemical additives in food should be listed by name on the label.

CHEMICALS AND ANIMALS

The use of antibiotics and hormones in feed should be reevaluated. Chances are, such practices should be abolished. Some control over the administration of medications to ill animals should be established. Since farmers are not trained in veterinary medicine, diagnosis and therapy should be left to veterinarians.

FROZEN FOODS

A complete survey of frozen foods, from the time of slaughter or harvesting to its preparation, packaging, sale, cooking, and appearance on the dining-room table, should be made.

No truck should be permitted to transport frozen food without adequate refrigeration and a graph to determine temperature control from the time of pickup to the time of delivery.

New safeguard devices, such as the strip of tape that shows whether a package has been defrosted, should be developed and used.

BETTER USE OF MANPOWER

There should be a coordinated inspection of food from growth and harvesting to the manner in which it is recommended that it be cooked. Overlapping and gaps in the use of manpower should be abolished.

Pay scales should be raised, and an effective, concerted effort should be made to interest young people in food-protection services.

FISH

Fish should be inspected under the law, and stricter standards should be made and enforced.

The American shipping industry should be revitalized by private and government subsidies for new equipment and by restrictions on imports.

SOME OBSERVATIONS

Our growing methods are not as efficient as those of the Japanese. If they were, California could feed the nation.

Our trucking is not as adequate as Britain's, where temperature graphs are required by law.

Our regulations governing pesticides are not as strict as Germany's.

Our food doesn't taste as good as French food does.

The food we eat on earth is not as safe as the food we provide for our astronauts, where upsetting the delicate balance between man and microbes could cause disaster.

But our supply is wondrous in its abundance and variety. Our sanitation is above that of countries where diarrhea kills off a large percentage of the child population.

For all-around quality, the American food supply is the best in the world—but it could be much better—and safer!

10 NOTES

1. *A Strategy for a Livable Environment,* a report to the Secretary of Health, Education, and Welfare by the Task Force on Environmental Health and Related Problems, June,1967.

1a. "Upper Income Infants Fed Poorly, Study Says," *Pediatric Herald,* December, 1969, p. 6.

1b. Robert B. Choate, Jr., testimony before the U.S. Senate's Consumer Subcommittee, July 22, 1970, Washington, D.C.

2. Paul LaChance, Ph.D., Associate Professor of Nutritional Physiology, Rutgers University.

3. United States Department of Health, Education, and Welfare, news release, July 31, 1965.

3a. *Agricultural Research,* "The Wheat We Eat," September, 1970, pp. 12–13.

3b. N. R. Di Luzio, "Diet May Need More Vitamin E, Less Polyunsaturates," The American Chemical Society, March 30, 1971, Los Angeles, California.

3c. Aloys Tapell, American Societies for Experimental Biology Meeting, April 16, 1970, Atlantic City, N.J.

4. News from the National Institutes of Health, August 1, 1968; University of Wisconsin news release, June 12, 1968.

5. *A Strategy for a Livable Environment, loc. cit.*

6. *Ibid.*

7. News from the National Institute of Health, June 15, 1968.

8. Richard Farrow, Frank Lamb, Edgar Elkins, Russel Cook, Margaret Kawai, Antoni Cortes, "Effect of Commercial and Home Preparative Procedures on Parathion and Carbaryl Residues in Broccoli," American Chemical Society Meeting, San Francisco, April 1–5, 1968.

9. National Institutes of Health *Record,* July 10, 1968.

10. Richard L. Woodward, Ph.D., Boston, in *Journal of the American Medical Association,* July 1, 1968, Vol. 205, No. 1. *See also* J. D. Orgeron, M.P.H.; J. D. Martin, M.D.; C. T. Caraway, D.V.M.; Rosemary Martine, B.S.; George Hauser, M.D., *Public Health Reports,* March, 1957, Vol. 72.

11. Herbert E. Hall, Ph.D.; Keith Lewis, Ph.D., in October, 1967, issue of *Health Laboratory Science,* American Public Health Association.

12. Correspondence with author, July 18, 1968.

13. *A Strategy for a Livable Environment, loc. cit.*

14. *A Study of State and Local Food and Drug Programs,* report to the Commissioner of the Food and Drug Administration, U.S. Department of Health, Education, and Welfare, 1965.

15. *New York Times,* January 27, 1968.

16. *An Evaluation of Public Health Hazards from Microbiological Contamination of Foods,* a report of the Food Protection Committee of the Food and Nutrition Board, National Academy of Sciences and National Research Council, Publication 1195, 1964.

17. *Medical Tribune,* March 21, 1968.

18. Correspondence with FDA, June 5, 1968.

19. Health Bulletin, March 9, 1968: correspondence with author, June 10, 1968.

20. James Goddard, M.D., Commissioner of the Food and Drug Administration, March and April speeches, 1967.

21. George Larrick, Commissioner of the Food and Drug Administration, speech before the National Association of Frozen Food Packers, Chicago, March 20, 1964.

22. John Morris, *New York Times,* March 31, 1968.

23. Joseph G. Minish (D–New Jersey), personal communication with author, May 25, 1971.

24. Weems Clevenger, District Director, FDA, New York, tape-recorded interview with author, July, 1968.

25. *FDA Papers,* April, 1968.

26. Hearings before a subcommittee of the Committee on Government Operations, House of Representatives, March 29, 1971, Washington, D.C.

27. Richard L. Woodward, Ph.D., Boston, in the *Journal of the American Medical Association,* July 1, 1968, Vol. 205, No. 1. Donald E. Crane, Malco Chemical Co., American Society of Agricultural Engineering, Athens, Ga., 1965. J. D. Orgeron *et al.,* "Methemoglobinemia from Eating Meat with High Nitrite Content," *Public Health Reports,* March, 1957.

11

Home Fires and
Freezers

No matter how well our food is grown, manufactured, inspected, warehoused, and retailed, if we don't handle and prepare it properly ourselves in our homes, we and those we love will suffer from foodborne illnesses.

Based on the advice of experts in all fields, here are methods for protecting the food you buy:

• Do your grocery shopping last, and head immediately for home.

• Always wash your hands before preparing food and after touching meat, and before touching produce that will not be cooked.

• Don't allow anyone with an infection, including yourself, to prepare food.

• Purchase—and use—an accurate thermometer in your refrigerator.

• Keep the inside of the refrigerator clean. Wash it at frequent intervals.

• Keep the areas around the motor and refrigerating unit clean. Lint and dirt on these parts cut off the supply of air, causing the motor and the refrigeration unit to overwork, thus reducing their efficiency.

• Check the gaskets around the doors. Be sure that they are flexible, and prevent the cold air from escaping.

• If your refrigerator is not self-defrosting, check the cooling area frequently, and defrost it. A buildup of ice on the cooling coils acts as an insulator, and the refrigerator will not work so well as it should. When the ice builds up to a fourth of an inch, it is time to defrost.

• Store foods in small, shallow containers. The more

surface that is exposed to the cold, the faster the food will cool. Keep foods covered so that the food particles from the shelf above will not fall into the food stored below. Allow space between food containers and between all foods and the walls of the box.

• Never cover the wire shelves of the box with paper or foil, as this cuts down on the circulation of air within the refrigerator. Never stack foods on top of each other.

• Because bacteria grow between 45 F. and 120 F., refrigerate food immediately after it is cooked. This means all foods, including roasts, stews, leftovers, broths, puddings and so on. Any delay will allow bacteria to grow more rapidly.

•If much hot food must be refrigerated, precool it first. Place the food in a shallow pan, and set it in cold water. Be careful, of course, not to let the water mix with the food. Unless the cooling unit is overloaded, it is perfectly safe to put hot things in today's modern refrigerator.

• Keep hot things hot. Use a meat thermometer to check pork and other meats. In general, foods should be kept hot at a temperature above 140 degrees F.

• If you have been delayed in getting newly purchased food home to your refrigerator by a traffic jam or some other incident, be sure that temperatures reach 165 to 170 degrees in the center of the food that is being cooked.

• Keep eggs in the refrigerator.

• Store bread, in its original wrapper, in a breadbox or refrigerator. Bread keeps its freshness longer at room temperature, but in hot, humid weather, it is better protected against mold in the refrigerator.

• Store flours, cereals, spices, and other grains at room temperature in tightly closed containers that keep out dust, moisture, and insects. During the summer, buy such foodstuffs in small quantities only.

• Store butter, fat drippings, and margarine in tightly wrapped or covered containers in the refrigerator. These products should be used within two weeks. Don't let them stand for long periods at room temperature.

• Keep all homemade salad dressing in the refrigerator. Purchased mayonnaise and other ready-made salad dressings should be refrigerated unless used within a few days.

• Cold cuts should be stored in the refrigerator and used within three to five days.

• Store ham, frankfurters, bacon, bologna, and smoked sausage in the refrigerator in their original packaging. Uncooked cured pork may be stored longer than fresh pork, but the fat will become rancid if it is held too long. Bacon should be eaten within a week for best quality, a half a ham in three to five days, a whole ham within a week. Ham slices should be wrapped tightly and used within a few days.

• Ground meats, such as hamburger and fresh bulk sausage, are more likely to spoil than roasts, chops, or steaks because they have been exposed to contamination from air, food handlers, and mechanical equipment. Store them loosely wrapped in the coldest part of the refrigerator, and use them within one or two days.

• Although some food scientists say you can keep leftover stuffing from cooked fowl separate from the bird in the refrigerator for one or two days, it is best to throw it out.

• .Fresh milk and cream should be stored in the refrigerator at about 40 degrees F. Milk and cream are best stored only three to five days. Keep them covered so they won't absorb odors and flavors from other foods.

Keep dry milk in a tightly closed container, and reconstituted dry milk in the refrigerator.

•Keep hard cheese in the refrigerator. Wrap it tightly to keep out air, and trim away any mold that forms on the surface of the cheese before use.

• Store soft cheeses in tightly covered containers in the coldest part of the refrigerator. Use cottage cheese within three to five days, and others within two weeks.

• Keep synthetic custards, milks, and puddings in the refrigerator.

• With few exceptions—such as potatoes, dry onions, hard-rind squashes, and eggplants—keep vegetables in the refrigerator. Discard any vegetables that are bruised or soft or that show evidence of decay or worm injury.

• Wash all fruits and vegetables before eating them.

• Store honey and syrups at room temperature until opened. After their containers have been opened, honey and syrups are better protected from mold in the refrigerator.

• Store nuts in airtight containers in the refrigerator. Because of their high fat content, nuts require refrigeration to delay rancidity.

• After a jar of peanut butter has been opened, it should be refrigerated.

• When freezing food, package it so no moisture can enter. Containers should be leakproof.

• Records should be kept so that you know what foods are on hand and what foods need replenishing in the freezer. This will keep you from storing food too long.

• If a power failure develops in your freezer, or if the source of power is temporarily shut down, keep your freezer closed to retain the low temperature. Frozen foods in a fully loaded freezer will often stay frozen for as long as two days—in a half-filled freezer for about a day.

• Thawed fruits can be safely refrozen if they taste and smell good.

• Other foods that have thawed spoil more quickly. Special care is necessary in handling vegetables, shellfish, and cooked foods that have thawed or partially thawed.

• Once a food has been thawed, it is wise to use it as soon as possible, and not to refreeze it.

• Do keep frozen foods, especially poultry, in the refrigerator during thawing. If you have to speed up the defrosting of a turkey, place it under running cold water. Roasts, including turkeys, may go into the oven frozen; just increase the roasting time.

• Do keep meringues, custard-filled eclairs, synthetic custards, doughnuts, and pastries in the refrigerator. Take them out just in time to serve.

• Do chill main-dish salads for large groups in shallow bowls or on trays in the refrigerator, and keep an eye on their temperature during the meal.

Do keep sandwiches or sandwich fillings in the refrigerator until served. Covering sandwiches with a damp cloth is not recommended, since bacteria grow especially well under such circumstances. In fact, sandwiches should not be made the night before. Cooked meat next to damp bread makes an excellent breeding place for bacteria.

• Do keep a separate pan for washing dishes. Dishwater should be 130 degrees F. or above. Be sure that the rinse water is clean and hot and that the towels are clean.

• Don't empty the turtle's bowl or clean the dog's dish in the same sink you use for washing dishes and preparing the family's food. The turtle's bowl is full of his waste, and your pet's dish may well be contaminated with harm-

ful bacteria. In fact, avoid allowing pets in areas where food is being prepared.

• Don't use or buy canned foods if any of the following signs are present: bulging of the top or bottom of the can, dents along the side seams, or any sign of seepage, off-odor, or foaming when the can is opened; any unusual milky quality of liquid.

• Don't buy or eat salads, especially mayonnaise-base salads, that have not been or are not refrigerated.

• Don't use leftover food if discoloration, off-odor, or mold is apparent. Any food that has not been refrigerated below 45 degrees may be considered to be slightly spoiled. Recook any leftover food that has been kept in the refrigerator thirty-six hours or longer.

The cardinal rule of food safety at home is: When in doubt, throw it out!

Remember that we consumers have the weakest lobby in Washington. Our own education is our strongest safeguard against poisons in our food.

12

How to Crack the Food Freshness Codes

Food processors and retailers have a variety of codes to inform retailers and distributors—but not customers—about the freshness of products. Food-sellers claim that if housewives were able to read the codes, they would ignore wholesome but "out-of-date" products and concentrate just on the freshest and therefore would raise the cost of food. There have been a number of bills in state and federal legislatures to force the open-dating of food. Some chains have already introduced open-dating. Among them are Jewel Food Stores, Chicago; Safeway, Washington, D.C.; Grand Union and Pathmark, both in New Jersey; and Stop and Shop in Boston.

Generally, one of two possible coded dates is on the product—either the date the product was processed or the date it should be "pulled" from the shelf. Generally highly perishable products are given pull dates. Some of the codes, however, will require that you take along a pocket-size calendar and a pencil.

Food processors are often masters at making codes inconspicuous. On frozen-food packages, the dates are usually indented on the wrapper or the carton. These colorless indentations are usually at one end of the package. Another method of dating frozen food involves putting a small letter or number on the food wrapper. It is not stamped, but is part of the printing on the wrapper. Cans have the code numbers embossed on one end of the can, usually the bottom. Boxes have either stamped or indented codes on one end of the package.

It would be impossible to give all the codes here, but

most supermarkets keep a thick book of master codes in their offices. If you have no success in breaking the code on a particular product, and store personnel feign ignorance, ask to see the code book. If the store refuses, write to your congressman and send carbon copies of your letter to your local newspaper, board of health, and to the manufacturer. Chances are, you will get quick action.

The following are examples of most of the codes used by food processors. With a little patience, you will be able to break almost all the codes on the foods you purchase.

MEAT

Many stores and processed meat packagers follow the dating code of the American Meat Institute, in which the second and third digits signify the day and the first and last digits denote the month. (Ground meat should be used in one or two days after grinding. Cured and ready-to-serve meats should be used within a week for top freshness.)

Example 1: 1294 1 + 4 = 5th month, 29 = 29th day: MAY 29
Among stores and manufacturers using the American Meat Institute code are Pathmark, Grand Union, Swift, Taylor, Vienna, Hy-Grade, Jones, Oscar Mayer, Plymouth Rock (all expiration or pull dates), and Zion Kosher Meat Company, Hormel, Weavers' (all manufactured or processed dates).

Other coding methods used for meat:

Example 2: 615 6 = 6th month, 15 = 15th day: JUNE 15
Grand Union and Hebrew National Kosher Foods, Inc., are among those using this code. Note that different packaging, as in the case of Grand Union here and above, gets a different code.

Example 3: 21 Day of current month: JANUARY 21
Example 4: 06/26/71 JUNE 26, 1971

Hebrew National carries this expiration-date code on its packaged meats.

Example 5: D041043 D = April (A = January), 04 = 4th, day, 1 = 1971, 043 = disregard: APRIL 4, 1971

Armour uses this seven-character code for nonrefrigerated canned meat.

DAIRY PRODUCTS

Many states and municipalities already require open-dating, and the consumer can easily see when milk is out of date. For most dairy products, the date, coded or open, is the "pull" date. Cream cheese, which for top quality should be used within two weeks, may be coded in a variety of ways, and the code includes the plant vat number or location number plus the date it should be pulled.

Cheese

Example 1: B1220 B = vat, 12 = 12th month, 20 = 20th day: DECEMBER 20

Example 2: Feb. 14 This cream cheese should be pulled from the shelves February 14.

Example 3: 234P1 234 = 234th day of the year (or July 19), P = plant, 1 = 1971: JULY 19, 1971

Kraft uses this code for its sliced cheese.

The following are all production-date codes for, in order, Swiss Knight cheese, Wispride cheese, Kraft grated cheese, and Progresso grated cheese.

Example 4: 10722 1 = 1971, 07 = 7th month, 22 = 22nd day: JULY 22, 1971

Example 5: 21217 212 = 212th day of the year (or July 31), 1 = 1971, 7 = disregard: JULY 31, 1971

Example 6: 233G9 233 = 233rd day of the year (or August 21), G9 = disregard: AUGUST 21

Example 7: 180/71 180 = 180th day of the year (or June 29), 71 = 1971: JUNE 29, 1971

Cottage Cheese, Yogurt, Sour Cream

Cottage cheese (which should be used within three to five days), yogurt and sour cream (both of which should be used within ten days) also have a variety of codes. Here are some of them:

Example 1: 0429   04 = 4th month, 29 = 29th day: APRIL 29
This is the pull date. Sealtest uses this code in their first four numbers; additional numbers are for internal use.

Example 2: 5A21   5 = lot number, A = January(B = February), 21 = 21st day: JANUARY 21
Grand Union and Borden use this code: Dairylea uses the letter first.

Example 3: 1304   1 + 4 = 5th month, 30 = 30th day: MAY 30
This is a common code. Among those using it are Crowley, Elmhurst, and Dairylea (on some products).

Example 4: 5050   5 = 5th month (1–6 = January–June; also July–December), 05 = 5th day, 0 = disregard: MAY 5
Dannon uses this as a pull date for its yogurt and Danny and Bokoo products.

Example 5: 0620   06 = 6th month, 20 = 20th day: JUNE 20
Light n'Lively, Breakstone, and Pathmark use this expiration-date code for yogurt.

Example 6: 5F20   5 = disregard, F = June (A = January), 20 = 20th day: JUNE 20
Borden uses this expiration-date code for its yogurt.

Whips

Example 1: A05   A = January (B = February), 05 = 5th day: JANUARY 5
Reddi wip and Pathmark carry this expiration date.

Example 2: 1251   1 = 1971, 251 = 251st day (or September 8): SEPTEMBER 8, 1971

Eggs

The dating of eggs usually depends on the largess of the retailer. However, if the retailer wants the eggs graded

according to the Department of Agriculture, the date must be stated in an understandable form. The expiration date may not exceed ten days from the date the eggs were packed, excluding the day of packing. However, for top quality, eggs should be used within a week after the hens produce them. Eggs are usually marked with the expiration date.

Example 1: September 8, 1971 or 9–8

Butter and Margarine

Butter usually carries an embossed code at one end of the bar or package. Salted butter should be removed from sale 30 to 45 days after cutting; sweet butter, after 14 days. For top quality, butter and margarine should be used within two weeks. One code frequently used includes the grading numbers and the date.

Example 1: 12341231   1234 = government trading code,
   12 = 12th month, 31 = 31st day: DECEMBER 31
Example 2: 1280452   128 = 128th day (or May 8), 0452 = disregard: MAY 8
Breakstone, Hotel Bar, and Land O'Lakes butter use this production-date code.

Example 3: J–15   J = October (A = January), 15 = 15th day: OCTOBER 15
Nucoa and Mazola margarines use this production-date code.

Example 4: J 1124AL   J = disregard, 1 = 1971, 1 = September (J = October), 24 = 24th day, AL = disregard: SEPTEMBER 24, 1971
Chiffon margarine uses this expiration-date code.

Example 5: 233G1   233 = 233rd day of the year (or August 21), G = disregard, 1 = 1971: AUGUST 21, 1971
Kraft-Parkay margarine use this production-date code.

Example 6: 11091   11 = 11th month, 09 = 9th day, 1 = 1971: NOVEMBER 9, 1971
Lever Brothers carries this production-date code for its Imperial, Good Luck, and Golden Glow margarines.

Example 7: 091213   09 = 9th month, 12 = 12th day, 13 = disregard:

SEPTEMBER 12
Mother's margarine carries this production-date code.

Example 8: 11-3   11 = 11th month, 3 = 3rd day: NOVEMBER 3
Standard Brands' Blue Bonnet and Fleischmanns margarines carry this expiration-date code.

Example 9: 0926   09 = 9th month, 26 = 26th day:
SEPTEMBER 26
Breakstone and Borden margarines carry this expiration-date code.

BAKERY PRODUCTS

Bakers use a variety of codes including letters, colored twist ties, lines, and imprinted dates. The date is a pull date. For maximum freshness, baked goods should be used within two or three days.

Example 1: F F = Saturday (A = Monday; there is no G as baking is usually not done on Sunday)
Among companies using letter codes for their breads are Thomas Bakers, Bond (doughnuts), Silvercup, and American.

Example 2: red twist tie = Tuesday
Each manufacturer has a different color code to make it easy for route men to spot bread loaves that need to be removed from the shelves. Among companies using twist-tie codes are Grand Union, Bond, Wonder Bread, Pepperidge Farm (butterfly rolls and Danish pastry), and Silvercup.

Example 3: = = Tuesday (Sunday would be a single line (−); Mondays no baking; and Saturday six lines

$$\left(\begin{matrix}=\\=\\=\end{matrix}\right)$$

Bread in wrappers with end seals have a variety of codes. Some, like Grand Union's, in this example, are coded with lines that show the day it is to come off sale. Many bakers

use two codes, perhaps a blue twist for Thursday and a coded date to show the day it was manufactured or should be pulled from the shelf.

Example 4: Bond and Hollywood Diet are color coded; Grossinger's rye is letter coded.

Example 5: Pillsbury carries the expiration date on its rolls and biscuits openly, such as Oct. 25.

Example 6: 10–13 10 = October (11 = November), 13 = 13th day: OCTOBER 13

Borden carries this expiration-date code.

GROCERIES

These include salad dressings, cake mixes, cookies and crackers, cereals, mayonnaise, piecrust, and frozen foods. Most of these products include the date they should be removed from the shelves.

Cake Mixes
(all production dates below)

Example 1: A119T A = June (B = July; L = May), 1 = 1971, 19 = 19th day, T = disregard: JUNE 19, 1971

General Mills, Betty Crocker, and Gold Medal are included here.

Example 2: 0271B 02 = 2nd month, 71 = 1971, B = disregard: FEBRUARY 1971

Duncan Hines is included here.

Example 3: B71 B = 2nd month, 71 = 1971: FEBRUARY 1971

Pillsbury is included here.

Example 4: A10993 A = disregard, 1 = 1971, 099 = 99th day of the year (or April 9), 3 = disregard: APRIL 9, 1971

Swans Down is included here.

Cookies and Crackers

Example 5: 1B103 1B = disregard, 103 = 103rd day of the year (or April 13): APRIL 13

Nabisco carries this production-date code.

Example 6:   Q6D   Q = May (M = January; X = December),
   6 = 6th day, D = plant (disregard): MAY 6
Sunshine carries this expiration-date code.

Example 7:   Frozen or refrigerated doughs. These are marked with
   open dates signifying the day they should no longer be used.

The following codes are for varieties of breakfast rings,
doughnuts, cakes, pies, potato chips, cheese snacks, and
pretzels. These are marked with a combination of letters
and numbers:

Example 8:   2H2   H = August (A = January, B = February, etc;
   I is omitted), 22 = 22nd day: AUGUST 22
Example 9:   J5   J = September, 5 = 5th day: SEPTEMBER 5

Cereals

Temperature, place of storage, and insect infestation all
affect the shelf life of cereals. Cereals are coded with the
production dates on the boxes.
Example 10:   2B2045A   2 = 2nd month, B = plant, 2 = 1972,
   045 = 45th day of the year (or February 14), A = internal use:
   FEBRUARY 14, 1972
Kellogg's uses this code.

Example 11:   2150   2 = 1972, 150 = 150th day of the year
   (or May 29): MAY 29, 1972
Post uses this code.

Example 12:   1A21   1 = 1971, A = January (B = February),
   21 = 21st day: JANUARY 21, 1971
H-O Oats uses this code.

Example 13:   1L17F   1 = 1971, L = disregard, 17 = 17th day,
   F = June (A = January): JUNE 17, 1971
Quaker Oats uses this code.

Example 14:   12   1 = 1971, 2 = 2nd month: FEBRUARY 1971
Ralston uses this code.

Example 15:   Code based on color of dot on package, top side.
   Sequence of colors: green, orange, black, red, blue, and yellow.
   Black should be sold before red, blue, and yellow.
General Mills cereals carry this code.

Example 16:   2C29   2 = 2nd month, C = 1972 (B = 1971)
   29 = 29th day: FEBRUARY 29, 1972

Grand Union uses this expiration-date code for various groceries.

Example 17: 1052   1 = 1971, 052 = the 52nd day of the year
(or February 21): FEBRUARY 21, 1971
This production-date code is used by Birds Eye frozen foods.

Example 18: ●A21   ● = odd year (no dot = even year),
A = January (B = February), 21 = 21st day:
FEBRUARY 21, 1971
Mazola oil uses this production-date code.

Appendix

Papain is a meat tenderizer commonly used in homes and restaurants. An enzyme, it causes basic changes to occur in the rabbit's body when it is injected, according to researchers of The National Institutes of Health, U.S. Department of Health, Education, and Welfare. These changes result in the progressive collapse of a rabbit's normally erect ears.

Ordinarily, cartilage restoration takes place rapidly, and the ears stand up straight in three to five days. It was discovered, however, that the ears would remain collapsed—indicating interference with cartilage recovery—if daily injections of steroid drugs were given. The reason Federal authorities have not been alarmed over papain's effect on cartilage is that the papain is deactivated by heat used in cooking.

However, should the papain be put on steak tartar—uncooked meat—which is eaten raw or for some other reason not deactivated, the possibility of damage to the human body—especially in arthritics and asthmatics who take steroids—is not remote.

Dr. James W. Holsinger, Jr., Dr. Robert L. Fuson, and Dr. Will C. Sealy, all of the Division of Thoracic Surgery, Duke University Medical Center, described a death resulting from papain ingestion in the May 20, 1968, *Journal of the American Medical Association,* in an article entitled "Esophageal Perforation Following Meat Impaction and Papain Ingestion":

Papain has been given to people who have meat impactions in their esophagus (food pipe) after improperly

chewing or bolting their food. In the case described, the meat, after being in place for a long time, caused necrosis or injury to the tissue of the esophageal wall. According to the physicians, the papain given to the patient "digested this area."

"Perhaps the products of protein digestion caused other areas of necrosis which in turn were digested by the papain," the doctors reasoned.

Earlier, experimental studies on fourteen dogs showed every animal receiving papain had some degree of damage to the esophageal wall, with reddening and thickening of the mucosal wall at the site of the lodged bolus. Advanced hemorrhagic pulmonary edema was found to be the cause of death in these animals.

The thirty-nine-year-old patient described by Dr. Holsinger and his group died of hemorrhaging. It was the second case reported in recent years of death after ingestion of papain, and thus raises the question of what effect papain may have on necrotic tissue in the mouth, throat, or stomach.

The experiments with the rabbits raise the question of the effect of papain when steroids are present either in the food eaten or in the person taking them therapeutically.

STATE AGENCIES ADMINISTERING CONSUMER FOOD PROTECTION ACTIVITIES*

Interest code—   (1) Fluid milk.   (6) Poultry.
other agency: (2) Dairy products.   (7) Eggs.
               (3) Meat.   (8) Fruits and vegetables.
               (4) Fish.   (9) Other foods.
               (5) Shellfish.   (10) Eating and drinking places.

Foods and associated processing and distribution

State	Principal Agency	Other agency	Interests of other agency
Alabama	Agriculture	Health	1, 3, 10
Alaska	Health	Agriculture	3, 8
Arizona	Health	Dairy Commission	2
		Livestock Sanitary Board	2, 3
		Agriculture and Horticulture Commission.	8
		Egg Inspector	7
Arkansas	Health	Plant Board	8
California	Health	Agriculture	1, 2, 3, 6, 7, 8, 9
Colorado	Health	Agriculture	2, 6, 7, 8, 9
Connecticut	Consumer Protection.	Agriculture	1, 8, 9
		Health	5
Delaware	Health	Agriculture	2, 6, 7, 8, 9
Florida	Agriculture	Health	1, 5, 10
		Hotel and Restaurant Commission.	10
Georgia	Agriculture	Health	1, 5, 10
Hawaii	Health	Agriculture	3, 6, 8
Idaho	Health	Agriculture	1, 2, 3, 6, 7, 8, 9
Illinois	Agriculture	Health	1, 10
Indiana	Health	University	6, 7, 8
Iowa	Agriculture	Health	(†)
Kansas	Health	Agriculture	2, 7, 8
		Health and Restaurant Board.	9, 10
Kentucky	Health	Agriculture	7
Louisiana	Health	Agriculture	6, 8, 9

* "Protecting Our Food," in *The Yearbook of Agriculture*, U.S. Government Printing Office, 1966.

Foods and associated processing and distribution

State	Principal agency	Other agency	Interests of other agency
Maine	Agriculture	Health	10
		Sea and Shore Fisheries	5
Maryland	Health	Agriculture	2, 7
Massachusetts	Health	Agriculture	1, 6, 7, 8
Michigan	Agriculture	Health	1
Minnesota	Agriculture	Health	10
Mississippi	Health	Agriculture	2, 3, 7, 8
Missouri	Health	Agriculture	2, 7, 9
Montana	Health	Livestock Sanitary Board	2, 6
		Agriculture	2, 3, 7, 8, 9
Nebraska	Agriculture	Health	(†)
Nevada	Health	Agriculture	7, 9,
New Hampshire	Health	Agriculture	6, 7, 8, 9
New Jersey	Health	Agriculture	6, 7, 8
New Mexico	Health	Agriculture	1, 7, 8, 9
New York	Agriculture	Health	1, 10
North Carolina	Agriculture	Health	10
North Dakota	Laboratories Commission.	Agriculture	1, 2, 9
		Health	1, 9, 10
		Livestock Sanitary Board	3
Ohio	Agriculture	Health	10
Oklahoma	Health	Agriculture	2, 3, 7, 9
Oregon	Agriculture	Health	10
Pennsylvania	Agriculture	Health	10
Rhode Island	Health	Agriculture	7, 8
South Carolina	Agriculture	Health	2, 9, 10
South Dakota	Agriculture	Health	10
Tennessee	Agriculture	Health, Conservation, and Commerce.	10
Texas	Health	Agriculture	8
Utah	Agriculture	Health	10
Vermont	Health	Agriculture	1, 7, 8, 9
Virginia	Agriculture	Health	1, 2, 4, 5, 10
Washington	Agriculture	Health	1, 4, 5, 10
West Virginia	Agriculture	Health	1, 9, 10
Wisconsin	Agriculture	Health	1, 10
Wyoming	Agriculture	Health	10

† No information available on specific interests.

Index

273

Heptachlor, soil dusting with, 29, 37

Herbicides, 15, 32

Hesperidin, 73

HEW, 73

Ho Man Kwok, Dr. Robert, on "Chinese food syndrome," 75

Holsinger, Dr. James W., on papain, 268–69

Hormonal "programming," altered by DDT, 27

Hormones: female estrogen, 61; hidden, as food additives, 61–63

Houser, Leroy, on fishing methods, 132

Hueper, Dr. W. C., on arsenicals, 60; on diethylstilbestrol, 61–62; on food dyes, 14; on pesticides, 32–33, 81–82

Hunt, E. G., on pesticides, 24

Hydrocarbons, chlorinated, 16; aldrin, 28, 30; DDT, 24, 32–35; dieldrin, 27, 30–31, 40; endrin, 29, 30–31; enzyme stimulating effect, 35; lindane, 21, 22, 29

Industries: canning, 150–51, 217; use of aliphatic and aromatic chlorinated hydrocarbons by, 33; waste from, 184

Insecticides: and blood abnormalities, 21; benzene hexachloride, 21; Captan, 37, 40, 78; causing acute granulocytic leukemia, 20; chlordane, 40; increasing reproduction in the female mosquito, 45; laboratory tests of, 36; Malathion, 15, 35; malformation caused by, 31, 32; orgaophosphate compounds, 35. See also Insecticides; Organophosphates

Insecticides, organophosphate, 36; chlorthion, 36; Malathion, 15, 36; parathion, 21, 36; phosdrin, 36; schradan, 36; TEPP, 36; thimet, 36

Insects: killing of, 20–47; pests, 15; use of pyrethrum to kill, 44

Inspectors: federal and local, 240–41; Food and Drug Administration, 12, 91, 206; reports of USDA, 103–04; sanitary, 196; United States Department of Agriculture, 15, 73–74, 105

International Congress of Gastroenterology, 17

Intestinal diseases, 17, 177

Intrauterine death, 17

Intrauterine development of ova, 15

Irradiation of food, 221–24

James, Dr. Douglas, on DDT, 26

Johnson, Paul E., on food packaging materials, 86

Journal of Agricultural and Food Chemistry, on N-nitrosamines and cancer, 120

Journal of Asthma Research, on allergies, 85

Journal of Diseases of Children, on frozen baby food, 219

Journal of Public Health, on processed foods, 212

Journal of the American Medical Association, on cancer caused by nitrates and nitrites, 66; on food sanitation, 196; on methemoglobinemia caused by nitrite, 183; on papain, 268–69

Journal of the American Veterinary Medical Association, on insecticides, feed additives, and therapeutic agents, 112

Journal of the New Jersey Medical Society, on pesticides, 28

Kelly, Dr. Margaret, on chemicals and cancer, 52

Kempe, Lloyd, research on botulism, 152

Kirk, J. Kenneth, on food sanitation, 120–21

Kosher sausage, unsanitary preparation of, 109

Kupchik, Dr. George: on monitoring of pesticides, 41; on supermarkets, 228

Kwalick, Dr. Donald S., on pesticides, 28

Laboratory: American Public Health Association, 40, 102, 190; animals, 92–93; Biological Sciences, Maryland, 134; Chesapeake Biological, of the Natural Resources Institute, 134; diagnosis of poultry disease, 112–13; examination of dried milk, 210; refrigerators for incubation of botulinuses, 152; studies on fish, 134–35; tests, 31–36

LaChance, Robert, on supermarkets, 227–28, 229

Lackey, Robert, on food handlers, 195, 201

Lancet, The, on carcinogens, 66; on organophosphates, 36

Larrick, George P., on frozen foods, 216, 243